Mathematics Manual for Water and Wastewater Treatment Plant Operators: Wastewater Treatment Operations

To properly operate a waterworks or wastewater treatment plant and to pass the examination for a waterworks/wastewater operator's license, it is necessary to know how to perform certain calculations. All operators, at all levels of licensure, need a basic understanding of arithmetic and problem-solving techniques to solve the problems they typically encounter in the workplace.

Hailed on its first publication as a masterly account written in an engaging, highly readable, user-friendly style, the fully updated *Mathematics Manual for Water and Wastewater Treatment Plant Operators: Wastewater Treatment Operations* covers all the necessary computations used in wastewater treatment today. It presents math operations that progressively advance to higher, more practical applications, including math operations that operators at the highest level of licensure would be expected to know and perform.

Features:

- Provides a strong foundation based on theoretical math concepts, which it then applies to solving practical problems for both water and wastewater operations.
- Updated throughout and with several new practical problems added.
- Provides illustrative examples for commonly used waterworks and wastewater treatment operations covering unit process operations found in today's treatment facilities.

Mathematics Manual for Water and Wastewater Treatment Plant Operators: Wastewater Treatment Operations

Math Concepts and Calculations

Third Edition

Frank R. Spellman

CRC Press is an imprint of the
Taylor & Francis Group, an **informa** business

Third edition published 2024
by CRC Press
6000 Broken Sound Parkway NW, Suite 300, Boca Raton, FL 33487–2742

and by CRC Press
4 Park Square, Milton Park, Abingdon, Oxon, OX14 4RN

CRC Press is an imprint of Taylor & Francis Group, LLC

First edition published by CRC Press 2004
Second edition published by CRC Press 2014

ISBN: 978-1-032-40689-3 (hbk)
ISBN: 978-1-032-40688-6 (pbk)
ISBN: 978-1-003-35431-4 (ebk)

DOI: 10.1201/9781003354314

Typeset in Times
by Apex CoVantage, LLC

Contents

Preface

This is volume 3 of an industry-wide bestseller hailed on its first publication as a masterly account written in an engaging, highly readable, user-friendly, show-and-display—a classic present-and-do—style of presentation. *Mathematics Manual for Water and Wastewater Treatment Plant Operators*, 3rd edition, volume 3, contains hundreds of worked examples presented in step-by-step training style; it is ideal for all levels of wastewater treatment operators in training and practitioners studying for advanced licensure. In addition, this manual is a handy desk reference and/or hand-held guide for daily use in making operational math computations.

This standard synthesis has not only been completely revised but also expanded from one to three volumes. Volume 1 covers basic math operators and operations, volume 2 covers computations commonly used in water treatment plant operations, and volume 3 covers computations used in wastewater treatment.

To properly operate a waterworks or wastewater treatment plant and to pass the examination for a waterworks/wastewater operator's license, it is necessary to know how to perform certain calculations. In reality, most of the calculations that operators at the lower level of licensure need to know how to perform are not difficult, but all operators need a basic understanding of arithmetic and problem-solving techniques to be able to solve the problems they typically encounter.

How about waterworks/wastewater treatment plant operators at the higher levels of licensure? Do they also need to be well-versed in mathematical operations? The short answer is absolutely. The long answer is that if you work in water or wastewater treatment and expect to have a successful career which includes advancement to the highest levels of licensure or certification (usually prerequisites for advancement to higher management levels), you must have knowledge of math at both the basic or fundamental level and advanced practical level. It is not possible to succeed in this field without the ability to perform mathematical operations.

Keep in mind that mathematics is a universal language. Mathematical symbols have the same meaning to people speaking many different languages throughout the world. The key to learning mathematics is to learn the language, symbols, definitions, and terms of mathematics that allow us to grasp the concepts necessary to solve equations.

In *Mathematics Manual for Water and Wastewater Treatment Plant Operators*, we begin by introducing and reviewing concepts critical to the qualified operators at the fundamental or entry level; however, this does not mean that these are the only math concepts that a competent operator needs to know to solve routine operation and maintenance problems. After building a strong foundation based on theoretical math concepts (the basic tools of mathematics, including fractions, decimals, percent, areas, volumes) in volume 1 and water operator math in volume 2, we move on to applied wastewater treatment math presented in this volume. Even though there is considerable crossover of basic math operations used by both waterworks and wastewater operators, we separate applied math problems for wastewater and water. We do this to aid operators of specific unit processes unique to waterworks and wastewater operations focused on their specific area of specialization.

What makes *Mathematics Manual for Water and Wastewater Treatment Plant Operators Volume 3* different from the other available math books available? Consider the following:

- The author has worked in and around water/wastewater treatment and taught water/wastewater math for several years at the apprenticeship level and at numerous short courses for operators.
- The author has sat at the table of licensure examination preparation boards to review, edit, and write state licensure exams.
- This step-by-step training manual provides concise, practical instruction in the math skills that operators must have to pass certification tests.
- The text is completely self-contained in three complete volumes. The advantage should be obvious—three separate texts with math basics and advanced operator math concepts contained in each allow the user to choose the proper volume for his or her use.
- The text is user-friendly; no matter the difficulty of the problem to be solved, each operation is explained in straightforward, plain English. Moreover, numerous example problems (several hundred) are presented to enhance the learning process.
- The first edition was highly successful and well-received, but like any flagship edition of any practical manual, there is always room for improvement. Thankfully, many users have provided constructive criticism, advices, and numerous suggestions. All these inputs from actual users have been incorporated into this new three-volume set.

To assure correlation to modern practice and design, we present illustrative problems in terms of commonly used in waterworks/wastewater treatment operations and associated parameters and cover typical math concepts for waterworks/wastewater treatment unit process operations found in today's waterworks/wastewater treatment facilities.

This text is accessible to those who have little or no experience in treatment plant math operations. Readers who work through the text systematically will be surprised at how easily they can acquire an understanding and skill in water/wastewater math concepts, adding another critical component to their professional knowledge.

A final point before beginning our discussion of math concepts: it can be said with some accuracy and certainty that without the ability to work basic math problems (i.e., those typical to water/wastewater treatment), candidates for licensure will find any attempts to successfully pass licensure exams a much more difficult proposition.

Frank R. Spellman
Norfolk, VA.

About the Author

Frank R. Spellman is a retired assistant professor of environmental health at Old Dominion University, Norfolk, Virginia, and the author of more than 160 books covering topics ranging from concentrated animal feeding operations (CAFOs) to all areas of environmental science and occupational health. He consults on homeland security vulnerability assessments for critical infrastructures, including water/wastewater facilities, and conducts audits for Occupational Safety and Health Administration and Environmental Protection Agency inspections throughout the country. Dr. Spellman lectures on sewage treatment, water treatment, and homeland security, as well as on safety topics, throughout the country and teaches water/wastewater operator short courses at Virginia Tech in Blacksburg.

1 Flow, Velocity, and Pumping Calculations

1.1 VELOCITY AND FLOW CALCULATIONS

$$\textbf{Velocity} = \frac{\textbf{Distance Traveled, ft}}{\textbf{Duration of Test, min}} = \textbf{ft / min}$$

Example 1.1

Problem:
A cork is placed in a channel and travels 400 ft in 2 min. What is the velocity of the wastewater in the channel in feet per minute?

Solution:

$$\text{Velocity} = \frac{\text{distance, ft}}{\text{Time}} = \frac{400\text{ ft}}{2\text{ min}} = 200\text{ ft/min}$$

Example 1.2

Problem:
A float travels 320 ft in a channel in 2 min and 16 sec. What is the velocity in the channel in feet per second?

Solution:

$$2\text{ min} + 16\text{ sec} = 136\text{ seconds total}$$

$$\text{Velocity} = \frac{320\text{ ft}}{136\text{ sec}} = 2.4\text{ ft/sec}$$

Example 1.3

Problem:
The distance between manhole #1 and manhole #2 is 110 ft. A fishing bobber is dropped into manhole #1 and enters manhole #2 in 32 sec. What is the velocity of the wastewater in the sewer in feet per minute?

Solution:

$$32\text{ sec} = 0.533\text{ min}$$

DOI: 10.1201/9781003354314-1

$$\text{Velocity} = \frac{110 \text{ ft}}{0.533 \text{ min}} = 206.4 \text{ ft/min}$$

$$\textbf{Flow in a Channel}\,(\mathbf{Q}) = (\mathbf{w})(\mathbf{d})(\mathbf{velocity})$$

Example 1.4

Problem:
A channel 48 in wide has water flowing to a depth of 2 ft. If the velocity of the water is 2.6 ft/sec, what is the flow in the channel in cubic feet per second?

Solution:

$$\text{Width} = 48 \text{ in.} = 4 \text{ ft}$$

$$\begin{aligned} Q &= (w)(d)(vel) \\ &= (4 \text{ ft}(2 \text{ ft})(2.6 \text{ ft/sec}) \\ &= 20.8 \text{ ft}^3/\text{sec} \end{aligned}$$

Example 1.5

Problem:
A channel 3 ft wide has water flowing to a depth of 2.2 ft. If the velocity through the channel is 110 ft/min, what is the flow rate in cubic feet per minute? In million gallons per day?

Solution:

$$Q = (3 \text{ ft})(2.2 \text{ ft})(110 \text{ ft/sec}) = 726 \text{ ft}^3/\text{min}$$

$$\frac{(726 \text{ ft}^3/\text{min})(1440)(7.48)}{1{,}000{,}000} = 7.82 \text{ MGD}$$

Example 1.6

Problem:
A channel 3 ft wide has water flowing at a velocity of 1.6 ft/sec. If the flow through the channel is 8.0 cu ft/sec, what is the depth of the water in the channel in feet?

Solution:

$$\begin{aligned} &Q = (w)(d)(vel) \\ &8.0 \text{ ft}^3/\text{sec} = (3 \text{ ft})(d \text{ ft})(1.6 \text{ ft/sec}) \\ &8.0 = (4.8)(d) \end{aligned}$$

$$\frac{8.0}{4.8} = d$$
$$1.77 \text{ ft} = d$$

Flow Through Full Pipe $Q = (0.785)(D, ft)(D, ft)(vel)$

Example 1.7

Problem:
The flow through a 2 ft diameter pipeline is moving at a velocity of 3.0 ft/sec. What is the flow rate in cubic feet per second?

Solution:

$$\begin{aligned} Q &= (0.785)(D, \text{ft})(D, \text{ft})(\text{Vel}) \\ &= (0.785)(2 \text{ ft})(2 \text{ ft})(3.0 \text{ ft/sec}) \\ &= 9.42 \text{ ft}^3/\text{sec} \end{aligned}$$

Example 1.8

Problem:
The flow through a 6 in diameter pipeline is moving at a velocity of 2.5 ft/sec. What is the flow rate in cubic feet per second?

Solution:

$$\begin{aligned} Q &= (0.785)(0.5 \text{ ft})(0.5 \text{ ft})(2.5 \text{ ft/sec}) \\ &= 0.49 \text{ ft}^3/\text{sec} \end{aligned}$$

Example 1.9

Problem:
An 8 in diameter pipeline has water flowing at a velocity of 3.1 ft/sec. What is the flow rate in gallons per minute?

Solution:

$$\begin{aligned} Q &= (0.785)(0.6667 \text{ ft})(0.6667 \text{ ft})(3.1 \text{ ft/sec}) \\ &= (1.082 \text{ ft}^3/\text{sec})(60)(7.48) = 486 \text{ gpm} \end{aligned}$$

Example 1.10

Problem:
The flow through a pipe is 0.8 cu ft/sec. If the velocity of the flow is 3.5 ft/sec and the pipe is flowing full, what is the diameter of the pipe in inches?

Solution:

$$Q = (0.785)(D, t)(D, ft)(vel)$$
$$0.8 = (0.785)(D^2)(3.5)$$
$$0.8 = (2.748)(D^2)$$
$$\frac{0.8}{2.748} = D^2$$
$$\sqrt{0.29} = D^2$$
$$0.54 \text{ ft} = D$$
$$6.48 \text{ in}$$

1.1.1 Flow through Pipe Flowing Less Than Full

Note:

Calculating the flow rate through a pipeline flowing less than full requires the use of a factor based on the ratio of water depth (*d*) to the pipe diameter (*D*), instead of using 0.785 in the area calculation. Calculate the *d/D* value, then use Table 1.1 to determine the factor to be used instead of 0.785.

Example 1.11

Problem:
A 12 in diameter pipeline has water flowing at a depth of 5 in. What is the gallons per minute flow if the velocity of the wastewater is 310 fpm?

Solution:

$$d/D = 5/12 = 0.42 = 0.3130 \text{ on Table 1.1}$$
$$Q = (0.3130)(1 \text{ ft})(1 \text{ ft})(310 \text{ ft/min})$$
$$= (97.03 \text{ ft}^3)(7.48)726 \text{ gpm}$$

Example 1.12

Problem:
A 10 in diameter pipeline has water flowing at a velocity of 3.3 fps. What is the gallons per day flow rate if the water is at a depth of 5 in?

Solution:

$$D = 10 \text{ in} = 0.8333 \text{ ft}$$
$$d = 5 \text{ in}$$

TABLE 1.1
Depth/Diameter Table

d/D	Factor	d/D	Factor	d/D	Factor	d/D	Factor
0.01	0.0013	0.26	0.1623	0.51	0.4027	0.76	0.6404
0.02	0.0037	0.27	0.1711	0.52	0.4127	0.77	0.6489
0.03	0.0069	0.28	0.1800	0.53	0.4227	0.78	0.6573
0.04	0.0105	0.29	0.1890	0.54	0.4327	0.79	0.6655
0.05	0.0147	0.30	0.1982	0.55	0.4426	0.80	0.6736
0.06	0.0192	0.31	0.2074	0.56	0.4526	0.81	0.6815
0.07	0.0242	0.32	0.2167	0.57	0.4625	0.82	0.6893
0.08	0.0294	0.33	0.2260	0.58	0.4724	0.83	0.6969
0.09	0.0350	0.34	0.2355	0.59	0.4822	0.84	0.7043
0.10	0.0409	0.35	0.2450	0.60	0.4920	0.85	0.7115
0.11	0.0470	0.36	0.2546	0.61	0.5018	0.86	0.7186
0.12	0.0534	0.37	0.2642	0.62	0.5115	0.87	0.7254
0.13	0.0600	0.38	0.2739	0.63	0.5212	0.88	0.7320
0.14	0.0668	0.39	0.2836	0.64	0.5308	0.89	0.7384
0.15	0.0739	0.40	0.2934	0.65	0.5404	0.90	0.7445
0.16	0.0811	0.41	0.3032	0.66	0.5499	0.91	0.7504
0.17	0.0885	0.42	0.3130	0.67	0.5594	0.92	0.7560
0.18	0.0961	0.43	0.3229	0.68	0.5687	0.93	0.7612
0.19	0.1039	0.44	0.3328	0.69	0.5780	0.94	0.7662
0.20	0.1118	0.45	0.3428	0.70	0.5872	0.95	0.7707
0.21	0.1199	0.46	0.3527	0.71	0.5964	0.96	0.7749
0.22	0.1281	0.47	0.3627	0.72	0.6054	0.97	0.7785
0.23	0.1365	0.48	0.3727	0.73	0.6143	0.98	0.7816
0.24	0.1449	0.49	0.3877	0.74	0.6231	0.99	0.7841
0.25	0.1535	0.50	0.3927	0.75	0.6318	1.00	0.7854

$$vel = 3.3 \text{ ft/sec}$$

$$d/D = 5/10 = 0.5 = 0.397 \text{ on Table 1.1}$$

$$\begin{aligned} Q &= (0.3927)(0.8333 \text{ ft})(0.8333 \text{ ft})(3.3 \text{ ft/sec}) \\ &= (0.8999 \text{ ft}^3/\text{sec})(60)(1440)(7.48) \\ &= 581{,}580 \text{ gpd} \end{aligned}$$

1.2 FLOW AND VELOCITY PRACTICE PROBLEMS

Problem 1.1

A channel is 3 ft wide, with water flowing to a depth of 2.2 ft. If the velocity in the channel is found to be 1.6 fps, what is the cubic feet per second flow rate in the channel?

Solution:

$$\begin{aligned} Q &= (w)(d)(vel) \\ &= (3 \text{ ft})(2.2 \text{ ft})(1.6 \text{ fps}) \\ &= 10.56 \text{ ft}^3/\text{sec} \end{aligned}$$

Problem 1.2

A 12 in diameter pipe is flowing full. What is the cubic feet per minute flow rate in the pipe if the velocity is 120 ft/min?

Solution:

$$\begin{aligned} Q &= (0.785)(D, \text{ft})(D, \text{ft})(vel) \\ &= (0.785)(1 \text{ ft})(1 \text{ ft})(120 \text{ ft/min}) \\ &= 94.2 \text{ ft}^3/\text{min} \end{aligned}$$

Problem 1.3

A water main with a diameter of 18 in is determined to have a velocity of 184 ft/min. What is the flow rate in gallons per minute?

Solution:

$$\begin{aligned} Q &= (0.785)(1.5 \text{ ft})(1.5 \text{ ft})(184 \text{ fpm}) \\ &= (324.99 \text{ ft}^3/\text{min})(7.48) = 2430.9 \text{ gpm} \end{aligned}$$

Problem 1.4

A 24 in main has a velocity of 206 ft/min. What is the gallons per day flow rate for the pipe?

Solution:

$$\begin{aligned} Q &= (0.785)(2 \text{ ft})(2 \text{ ft})(206 \text{ ft/min}) \\ &= (646.84 \text{ ft}^3/\text{min})(1440)(7.48) \\ &= 6{,}967{,}243 \text{ gpd} \end{aligned}$$

Problem 1.5

What would be the gallons per day flow rate for a 6 in line flowing at 2.1 ft/sec?

Solution:

$$\begin{aligned} Q &= (0.785)(0.5 \text{ ft})(0.5 \text{ ft})(2.1 \text{ ft/sec}) \\ &= (0.4121 \text{ ft}^3/\text{sec})(60)(1440)(7.48) \\ &= 266{,}329 \text{ gpd} \end{aligned}$$

Problem 1.6

A 36 in sewer needs to be cleaned. If the line is flushed at 2.2 ft/sec, how many gallons per minute of water should be flushed from the hydrant?

Solution:

$$\begin{aligned} Q &= (0.785)(3\text{ ft})(3\text{ ft})(2.2\text{ fps}) \\ &= (15.54\text{ ft/sec})(60)(7.48) \\ &= 6974\text{ gpm} \end{aligned}$$

Problem 1.7

A 36 in pipe has just been installed. If the wastewater is flowing at a velocity of 2.2 ft/sec, how many million gallons per day will the pipe deliver?

Solution:

$$\begin{aligned} Q &= (0.785)(3\text{ ft})(3\text{ ft})(2.2\text{ ft/sec}) \\ &= \frac{(15.54\text{ ft}^3\text{/sec})(60)(1440)(7.48)}{1{,}000{,}000} \\ &= 10.04\text{ MGD} \end{aligned}$$

Problem 1.8

A pipe has a diameter of 18 in. If the pipe is flowing full and the water is known to flow a distance of 850 yd in 5 min, what is the million gallons per day flow rate for the pipe?

Solution:

$D = 18\text{ in} = 1.5\text{ ft}$

Distance = (850 yd)(3 ft/yd) = 2,550 ft

Time = 5 min

Velocity = 2,550 ft/5 min = 510 ft/min

$$\begin{aligned} Q &= (0.785)(1.5\text{ ft})(1.5\text{ ft})(510\text{ ft/min}) \\ &= \frac{(900.8\text{ ft}^3\text{/min})(1440)(7.48)}{1{,}000{,}000} \\ &= 9.7\text{ MGD} \end{aligned}$$

Problem 1.9

A float is placed in an open channel. It takes 2.6 min to travel 310 ft. What is the flow velocity in feet per minute in the channel? (Assume that the float is traveling at the average velocity of the water.)

Solution:

$$\text{Velocity} = \text{Distance/time} = 310\ \text{ft}/2.6\ \text{min} = 119.2\ \text{ft/min}$$

Problem 1.10

A cork is placed in an open channel that is 30 ft in 22 sec. What is the velocity of the cork in feet per second?

Solution:

$$\text{Velocity} = 30\ \text{ft}/22\ \text{seconds} = 1.36\ \text{ft/sec}$$

1.3 PUMPING

Pumps and pumping calculations were discussed in detail in volumes 1 and 2 and are discussed in volume 3 because they are germane to many wastewater treatment processes and especially to their influent and effluent operations. Pumping facilities and appurtenances are required wherever gravity can't be used to supply water to the distribution system under sufficient pressure to meet all service demands. Pumps used in water and wastewater treatment are the same. Because the pump is so perfectly suited to the tasks it performs, and because the principles that make the pump work are physically fundamental, the idea that any new device would ever replace the pump is difficult to imagine. The pump is the workhorse of water/wastewater operations. Simply, pumps use energy to keep water and wastewater moving. To operate a pump efficiently, the operator and/or maintenance operator must be familiar with several basic principles of hydraulics. In addition, to operate various unit processes, in both water and wastewater operations, at optimum levels, operators should know how to perform basic pumping calculations.

1.4 BASIC WATER HYDRAULICS CALCULATIONS

1.4.1 Weight of Water

Because water must be stored and/or kept moving in water supplies and wastewater must be collected, processed, and discharged (out-falled) to its receiving body, we must consider some basic relationships in the weight of water; 1 cu ft of water weighs 62.4 lb and contains 7.48 gal, and 1 cu in of water weighs 0.0362 lb. Water 1 ft deep will exert a pressure of 0.43 lb/sq in on the bottom area (12 in × 0.062 lb/in^3). A column of water 2 ft high exerts 0.86 psi, one 10 ft high exerts 4.3 psi, and one 52 ft high exerts:

$$52\ \text{ft} \times 0.43\ \text{psi/ft} = 22.36\ \text{psi}$$

A column of water 2.31 ft high will exert 1 psi. To produce a pressure of 40 psi requires a water column of:

$$40\ \text{psi} \times 2.31\ \text{ft/psi} = 92.4$$

The term *head* is used to designate water pressure in terms of the height of a column of water in feet. For example, a 10 ft column of water exerts 4.3 psi. This can be called 4.3 psi pressure or 10 ft of head. If the static pressure in a pipe leading from an elevated water storage tank is 37 lb/sq in (psi), what is the elevation of the water above the pressure gauge?

Remembering that 1 psi = 2.31 and that the pressure at the gauge is 37 psi:

$$37 \text{ psi} \times 2.31 \text{ ft/psi} = 85.5 \text{ ft (rounded)}$$

1.4.2 Weight of Water Related to the Weight of Air

The theoretical atmospheric pressure at sea level (14.7 psi) will support a column of water 34 ft high:

$$14.7 \text{ psi} \times 2.31 \text{ ft/psi} = 33.957 \text{ ft or } 34 \text{ ft}$$

At an elevation of 1 mi above sea level, where the atmospheric pressure is 12 psi, the column of water would be only 28 ft high:

$$\left(12 \text{ psi} \times 2.31 \text{ ft/psi} = 27.72 \text{ ft or } 28 \text{ ft}\right).$$

If a tube is placed in a body of water at sea level (a glass, a bucket, a water storage reservoir, or a lake, pool, etc.), water will rise in the tube to the same height as the water outside the tube. The atmospheric pressure of 14.7 psi will push down equally on the water surface inside and outside the tube. However, if the top of the tube is tightly capped and all the air is removed from the sealed tube above the water surface, forming a *perfect vacuum*, the pressure on the water surface inside the tube will be 0 psi. The atmospheric pressure of 14.7 psi on the outside of the tube will push the water up into the tube until the weight of the water exerts the same 14.7 psi pressure at a point in the tube even with the water surface outside the tube. The water will rise 14.7 psi × 2.31 ft/psi = 34 ft.

In practice, it is impossible to create a perfect vacuum, so the water will rise somewhat less than 34 ft; the distance it rises depends on the amount of vacuum created.

Example 1.13

Problem:
If enough air was removed from the tube to produce an air pressure of 9.5 psi above the water in the tube, how far will the water rise in the tube?

Solution:
To maintain the 14.7 psi at the outside water surface level, the water in the tube must produce a pressure of 14.7 psi – 9.5 psi = 5.2 psi. The height of the column of water that will produce 5.2 psi is:

$$5.2 \text{ psi} \times 2.31 \text{ ft/psi} = 12 \text{ ft}$$

1.4.3 Water at Rest

Stevin's Law states, "The pressure at any point in a fluid at rest depends on the distance measured vertically to the free surface and the density of the fluid." Stated as a formula, this becomes:

$$p = w \times h \tag{1.1}$$

Where:

p = pressure in pounds per square foot (psf)
w = density in pounds per cubic foot (lb/cu ft)
h = vertical distance in feet

Example 1.14

Problem:
What is the pressure at a point 16 ft below the surface of a reservoir?

Solution:
To calculate this, we must know that the density of water w is 62.4 lb/cu ft. Thus:

$$\begin{aligned} p &= w \times h \\ &= 62.4\ \text{lb/ft}^3 \times 16\ \text{ft} \\ &= 998.4\ \text{lb/ft}^2 \text{or } 998.4\ \text{psf} \end{aligned}$$

Waterworks operators generally measure pressure in pounds per square **inch** rather than pounds per square **foot**; to convert, divide by 144 in^2/ft^2 (12 in × 12 in = 144 in^2):

$$p = \frac{998.4\ \text{lb/ft}^2}{144\ \text{in}^2/\text{ft}^2} = 6.93\ \text{lb/in}^2 \text{or psi}$$

1.4.4 Gauge Pressure

We have defined *head* as the height a column of water would rise due to the pressure at its base. We demonstrated that a perfect vacuum plus atmospheric pressure of 14.7 psi would lift the water 34 ft. If we now open the top of the sealed tube to the atmosphere and enclose the reservoir then increase the pressure in the reservoir, the water will again rise in the tube. Because atmospheric pressure is essentially universal, we usually ignore the first 14.7 psi of actual pressure measurements and measure only the difference between the water pressure and the atmospheric pressure; we call this *gauge pressure.*

Example 1.15

Problem:
Water in an open reservoir is subjected to the 14.7 psi of atmospheric pressure, but subtracting this 14.7 psi leaves a gauge pressure of 0 psi. This shows that the water

would rise 0 ft above the reservoir surface. If the gauge pressure in a water main is 110 psi, how far would the water rise in a tube connected to the main?

Solution:

$$110 \text{ psi} \times 2.31 \text{ ft/psi} = 254.1 \text{ ft}$$

1.4.5 Water in Motion

The study of water in motion is much more complicated than that of water at rest. It is important to understand these principles because the water/wastewater in a treatment plant and/or distribution/collection system is nearly always in motion (much of this motion is the result of pumping, of course).

1.4.5.1 Discharge

Discharge is the quantity of water passing a given point in a pipe or channel during a given period of time. It can be calculated by the formula:

$$Q = V \times A \tag{1.2}$$

Where:
Q = discharge in cubic feet per second (cfs)
V = water velocity in feet per second (fps or ft/sec)
A = cross-section area of the pipe or channel in square feet (ft^2)

Discharge can be converted from cubic feet per second to other units, such as gallons per minute (gpm) or million gallons per day (MGD), by using appropriate conversion factors.

Example 1.16

Problem:
A pipe 12 in in diameter has water flowing through it at 12 ft/sec. What is the discharge in (a) cubic feet per second, (b) gallons per minute, and (c) million gallons per day?

Solution:
Before we can use the basic formula, we must determine the area (A) of the pipe. The formula for the area is:

$$A = \pi \times \frac{D^2}{4} = \pi \times r^2$$

(π is the constant value 3.14159.)

Where:
D = diameter of the circle in feet
r = radius of the circle in feet

So the area of the pipe is:

$$A = \pi \times \frac{D^2}{4} = 3.14159 = 0.785 \text{ ft}^2$$

Now we can determine the discharge in cubic feet per second (part [a]):

$$Q = V \times A = 12 \text{ ft/sec} \times 0.785 \text{ ft}^2 = 9.42 \text{ ft}^3\text{/sec or cfs}$$

For part (b), we need to know that 1 cu ft/sec is 449 gal per minute, so 7.85 cfs × 449 gpm/cfs = 3,520 gpm.

Finally, for part (c), 1 MGD is 1.55 cfs, so:

$$\frac{7.85 \text{ cfs}}{1.55 \text{ cfs/MGD}} = 5.06 \text{ MGD}$$

THE LAW OF CONTINUITY

The *law of continuity* states that the discharge at each point in a pipe or channel is the same as the discharge at any other point (provided water does not leave or enter the pipe or channel). In equation form, this becomes:

$$Q_1 = Q_2 \text{ or } A_1V_1 = A_2V_2 \tag{1.3}$$

Example 1.17

Problem:

A pipe 12 in diameter is connected to a 6 in diameter pipe. The velocity of the water in the 12 in pipe is 3 fps. What is the velocity in the 6 in pipe? Using the equation $A_1V_1 = A_2V_2$, we need to determine the area of each pipe.

$$\begin{aligned} 12-\text{inch pipe}: A &= \pi \times \frac{D^2}{4} \\ &= 3.1419 \times \frac{(1 \text{ ft})^2}{4} \\ &= 0.785 \text{ ft}^2 \end{aligned}$$

$$\begin{aligned} 6-\text{inch pipe}: A &= 3.14159 \times \frac{(0.5)^2}{4} \\ &= 0.196 \text{ ft}^2 \end{aligned}$$

The continuity equation now becomes:

$$(0.785 \text{ ft}^2) \times (3 \text{ ft/sec}) = (0.196 \text{ ft}^2) \times V_2$$

Solving for V_2:

$$V_2 = \frac{(0.785\ \text{ft}^2) \times 3\ \text{ft/sec})}{(0.196\ \text{ft}^2)}$$

$$= 12\ \text{ft/sec or fps}$$

1.4.6 Pipe Friction

The flow of water in pipes is caused by the pressure applied behind it either by gravity or by hydraulic machines (pumps). The flow is retarded by the friction of the water against the inside of the pipe. The resistance of flow offered by this friction depends on the size (diameter) of the pipe, the roughness of the pipe wall, and the number and type of fittings (bends, valves, etc.) along the pipe. It also depends on the speed of the water through the pipe—the more water you try to pump through a pipe, the more pressure it will take to overcome the friction. The resistance can be expressed in terms of the additional pressure needed to push the water through the pipe, in either psi or feet of head. Because it is a reduction in pressure, it is often referred to as *friction loss* or *head loss.*

FRICTION LOSS INCREASES AS:

- Flow rate increases
- Pipe diameter decreases
- Pipe interior becomes rougher
- Pipe length increases
- Pipe is constricted
- Bends, fittings, and valves are added

The actual calculation of friction loss is beyond the scope of this text. Many published tables give the friction loss in different types and diameters of pipe and standard fittings. What is more important here is recognition of the loss of pressure or head due to the friction of water flowing through a pipe. One of the factors in friction loss is the roughness of the pipe wall. A number called the C factor indicates pipe wall roughness; the **higher** the C factor, the **smoother** the pipe.

Note: C factor is derived from the letter *C* in the Hazen–Williams equation for calculating water flow through a pipe.

Some of the roughness in the pipe will be due to the material; a cast iron pipe will be rougher than plastic, for example. Additionally, the roughness will increase with corrosion of the pipe material and deposits of sediments in the pipe. New water pipes should have a C factor of 100 or more; older pipes can have C factors very much lower than this. To determine C factor, we usually use published tables. In addition, when the friction losses for fittings are factored in, other published tables are available to make the proper determinations. It is standard practice to calculate the head loss from fittings by substituting the *equivalent length of pipe*, which is also available from published tables.

1.5 BASIC PUMPING CALCULATIONS

Certain computations used for determining various pumping parameters are important to the water/wastewater operator. In this section, we cover those basic pumping calculations important to the subject matter.

1.5.1 Pumping Rates

Important Point: The rate of flow produced by a pump is expressed as the volume of water pumped during a given period.

The mathematical problems most often encountered by water/wastewater operators in regard to determining pumping rates are often determined by using equations 1.4 and/or 1.5.

$$\text{Pumping Rate, }(\text{gpm}) = \frac{\text{gallons}}{\text{minutes}} \tag{1.4}$$

$$\textit{Pumping Rate, }(gph) = \frac{\text{gallons}}{\text{hours}} \tag{1.5}$$

Example 1.18

Problem:

The meter on the discharge side of the pump reads in hundreds of gallons. If the meter shows a reading of 110 at 2:00 p.m. and 320 at 2:30 p.m., what is the pumping rate expressed in gallons per minute?

Solution:

The problem asks for pumping rate in gallons per minute (gpm), so we use equation 14.4.

$$\text{Pumping Rate, gpm} = \frac{\text{gallons}}{\text{minutes}}$$

Step 1: To solve this problem, we must first find the total gallons pumped (determined from the meter readings).

$$\begin{array}{r} 32{,}000 \text{ gallons} \\ \underline{-11{,}000 \text{ gallons}} \\ 21{,}000 \text{ gallons} \end{array}$$

Step 2: The volume was pumped between 2:00 p.m. and 2:30 p.m., for a total of 30 min. From this information, calculate the gallons per minute pumping rate:

$$\begin{aligned} \text{Pumping rate, gpm} &= \frac{21{,}000 \text{ gal}}{30 \text{ min}} \\ &= 700 \text{ gpm pumping rate} \end{aligned}$$

Example 1.19

Problem:
During a 15 min pumping test, 16,000 gal were pumped into an empty rectangular tank. What is the pumping rate in gallons per minute?

Solution:
The problem asks for the pumping rate in gallons per minute, so again we use equation 1.4.

$$\begin{aligned}\text{Pumping rate, gpm} &= \frac{\text{gallons}}{\text{minutes}}\\ &= \frac{16{,}000 \text{ gallons}}{15 \text{ minutes}}\\ &= 1{,}067 \text{ gpm pumping rate}\end{aligned}$$

Example 1.20

Problem:
A tank 50 ft in diameter is filled with water to a depth of 4 ft. To conduct a pumping test, the outlet valve to the tank is closed and the pump is allowed to discharge into the tank. After 60 min, the water level is 5.5 ft. What is the pumping rate in gallons per minute?

Solution:
Step 1: We must first determine the volume pumped in cubic feet:

$$\begin{aligned}\text{Volume pumped} &= (\text{area of circle})(\text{depth})\\ &= (0.785)\ (50 \text{ ft})\ (50 \text{ ft})\ (1.5 \text{ ft})\\ &= 2{,}944 \text{ ft}^3 \ (\text{rounded})\end{aligned}$$

Step 2: Convert the cubic feet volume to gallons:

$$\left(2{,}944 \text{ ft}^3\right)\ \left(7.48 \text{ gal/ft}^3\right) = 22{,}021 \text{ gallons } (\text{rounded})$$

Step 3: The pumping test was conducted over a period of 60 min. Using equation 1.4, calculate the pumping rate in gallons per minute:

$$\begin{aligned}\text{pumping rate} &= \frac{\text{gallons}}{\text{minutes}}\\ &= \frac{22{,}021 \text{ gallons}}{60 \text{ minutes}}\\ &= 267 \text{ gpm } (\text{rounded})\end{aligned}$$

1.5.2 Calculating Head Loss

Important Note: Pump head measurements are used to determine the amount of energy a pump can or must impart to the water; they are measured in feet.

One of the principle calculations in pumping problems is used to determine *head loss*. The following formula is used to calculate head loss:

$$H_f = K(V^2/2g) \tag{1.6}$$

Where:

H_f = friction head
K = friction coefficient
V = velocity in pipe
g = gravity (32.17 ft/sec/sec)

1.5.3 Calculating Head

For centrifugal pumps and positive displacement pumps, several other important formulae are used in determining *head*. In centrifugal pump calculations, the conversion of the discharge pressure to the discharge head is the norm. Positive displacement pump calculations often leave given pressures in psi. In the following formulae, *W* expresses the specific weight of liquid in pounds per cubic foot. For water at 68°F, *W* is 62.4 lb/cu ft. A water column 2.31 ft high exerts a pressure of 1 psi on 64°F water. Use the following formulae to convert discharge pressure in pounds per square gauge to head in feet:

- Centrifugal pumps

$$\text{H, ft} = \frac{\text{P, psig} \times 2.31}{\text{specific gravity}} \tag{1.7}$$

- Positive displacement pumps

$$\text{H, ft} = \frac{\text{P, psig} \times 144}{\text{W}} \tag{1.8}$$

To convert head into pressure:

- Centrifugal pumps

$$\text{P, psi} = \frac{\text{H, ft} \times \text{specific gravity}}{2.31} \tag{1.9}$$

- Positive displacement pumps

$$\text{P, psi} = \frac{\text{H, ft} \times \text{W}}{\text{W}} \tag{1.10}$$

1.5.4 Calculating Horsepower and Efficiency

When considering work being done, we consider the "rate" at which work is being done. This is called *power* and is labeled as **foot-pounds/second**. At some point in the past, it was determined that the ideal work animal, the horse, could move 550 lb 1 ft in 1 sec. Because large amounts of work are also to be considered, this unit became known as *horsepower*. When pushing a certain amount of water at a given pressure, the pump performs work. One horsepower equals 33,000 ft-lb/min. The two basic terms for horsepower are:

- Hydraulic horsepower (whp)
- Brake horsepower (bhp)

1.5.4.1 Hydraulic Horsepower (WHP)

One hydraulic horsepower equals the following:

- 550 ft-lb/sec
- 33,000 ft lb/min
- 2,545 British thermal units per hour (Btu/hr)
- 0.746 kw
- 1.014 metric hp

To calculate the hydraulic horsepower (WHP) using flow in gallons per minute and head in feet, use the following formula for centrifugal pumps:

$$\text{WHP} = \frac{\text{flow, gpm} \times \text{hear, ft} \times \text{specific gravity}}{3{,}960} \tag{1.11}$$

When calculating horsepower for positive displacement pumps, common practice is to use pounds per square inch for pressure. Then the hydraulic horsepower becomes:

$$\text{WHP} = \frac{\text{flow, gpm} \times \text{pressure, psi}}{3{,}960} \tag{1.12}$$

1.5.4.2 Pump Efficiency and Brake Horsepower (BHP)

When a motor pump combination is used (for any purpose), neither the pump nor the motor will be 100% efficient. Simply, not all the power supplied by the motor to the pump (called *brake horsepower*, bhp) will be used to lift the water (*water or hydraulic horsepower*)—some of the power is used to overcome friction within the pump. Similarly, not all the power of the electric current driving the motor (called *motor horsepower*, mhp) will be used to drive the pump—some of the current is used to overcome friction within the motor, and some current is lost in the conversion of electrical energy to mechanical power.

Note: Depending on size and type, pumps are usually 50–85% efficient, and motors are usually 80–95% efficient. The efficiency of a particular motor or pump is given in the manufacturer's technical manual accompanying the unit.

The *brake horsepower (bhp)* of a pump is equal to hydraulic horsepower divided by the pump's efficiency. Thus, the bhp formula becomes:

$$\text{bhp} = \frac{\text{flow, gpm} \times \text{head, ft} \times \text{specific gravity}}{3{,}960 \times \text{efficiency}} \tag{1.13}$$

or

$$\text{bhp} = \frac{\text{flow, gpm} \times \text{pressure, psig}}{1{,}714 \times \text{efficiency}} \tag{1.14}$$

Example 1.21

Problem:

Calculate the brake horsepower requirements for a pump handling salt water and having a flow of 700 gpm with 40 psi differential pressure. The specific gravity of salt water at 68°F equals 1.03. The pump efficiency is 85%.

Solution:

To use equation 1.13, convert the pressure differential to total differential head, TDH = 40 × 2.31/1.03 = 90 ft.

$$\text{bhp} = \frac{700 \times 90 \times 1.03}{3{,}960 \times 0.85}$$

$$= 19.3 \text{ hp (rounded)}$$

Using equation 1.14:

$$\text{bhp} = \frac{700 \times 40}{1{,}714 \times 0.85}$$

$$= 19.2 \text{ hp (rounded)}$$

Important Point: Horsepower requirements vary with flow. Generally, if the flow is greater, the horsepower required to move the water would be greater.

When the motor, brake, and motor horsepower are known and the **efficiency** is unknown, a calculation to determine motor or pump efficiency must be done. Equation 14.15 is used to determine percent efficiency.

$$\text{Percent Efficiency} = \frac{\text{hp output}}{\text{hp input}} \times 100 \tag{1.15}$$

From equation 1.15, the specific equations to be used for motor, pump, and overall efficiency equations are:

$$\text{Percent Motor Efficiency} = \frac{\text{bhp}}{\text{mhp}} \times 100$$

$$\text{Percent Pump Efficiency} = \frac{\text{whp}}{\text{bhp}} \times 100$$

$$\text{Percent Overall Efficiency} = \frac{\text{whp}}{\text{mhp}} \times 100$$

Example 1.22

Problem:
A pump has a water horsepower requirement of 8.5 whp. If the motor supplies the pump with 10 hp, what is the efficiency of the pump?

Solution:

$$\begin{aligned}
\text{Percent pump efficiency} &= \frac{\text{whp output}}{\text{bhp supplied}} \times 100 \\
&= \frac{8.5 \text{ whp}}{10 \text{ bhp}} \times 100 \\
&= 0.85 \times 100 \\
&= 85\%
\end{aligned}$$

Example 1.23

Problem:
What is the efficiency if an electric power equivalent to 25 hp is supplied to the motor and 16 hp of work is accomplished by the pump?

Solution:
Calculate the percent of overall efficiency:

$$\begin{aligned}
\text{percent overall efficiency} &= \frac{\text{hp output}}{\text{hp supplied}} \times 100 \\
&= \frac{16 \text{ whp}}{25 \text{ mhp}} \times 100 \\
&= 0.64 \times 100 \\
&= 64\%
\end{aligned}$$

Example 1.24

Problem:
12 kW (kilowatts) of power is supplied to the motor. If the brake horsepower is 12 hp, what is the efficiency of the motor?

Solution:
First, convert the kilowatts power to horsepower. Based on the fact that 1 hp = 0.746 Kw, the equation becomes:

$$\frac{12 \text{ kW}}{0.746 \text{ kW/hp}} = 16.09 \text{ hp}$$

Now, calculate the percent efficiency of the motor:

$$\text{Percent efficiency} = \frac{\text{hp output}}{\text{hp supplied}} \times 100$$

$$= \frac{12 \text{ bhp}}{16.09 \text{ mhp}} \times 100$$

$$= 75\%$$

1.5.5 Specific Speed

Specific speed (N_s) refers to an impeller's speed when pumping 1 gpm of liquid at a differential head of 1 ft. Use the following equation for specific speed, where *H* is at the best efficiency point:

$$N_s = \frac{\text{rpm} \times Q^{0.5}}{H^{0.75}} \tag{1.16}$$

Where:

rpm = revolutions per minute
Q = flow (in gallons per minute)
H = head (in feet)

Pump specific speeds vary between pumps. No absolute rule sets the specific speed for different kinds of centrifugal pumps. However, the following N_s ranges are quite common.

- Volute, diffuser, and vertical turbine = 500–5,000
- Mixed flow = 5,000–10,000
- Propeller pumps = 9,000–15,000

Important Note: The higher the specific speed of a pump, the higher its efficiency.

1.6 POSITIVE DISPLACEMENT PUMPS

The clearest differentiation between centrifugal (or kinetic) pumps and positive displacement pumps is made or based on the method by which pumping energy is transmitted to the liquid. Kinetic (centrifugal pumps) relies on a transformation of kinetic energy to static pressure. Positive displacement pumps, on the other hand, discharge a given volume for each stroke or revolution (i.e., energy is added intermittently to the fluid flow). The two most common forms of positive displacement pumps are reciprocating action pumps (which use pistons, plungers, diaphragms, or bellows) and rotary action pumps (using vanes, screws, lobes, or progressing cavities). Regardless of form used, all positive displacement pumps act to force liquid into a system regardless of the resistance that may oppose the transfer. The discharge pressure generated by a positive displacement pump is, in theory, infinite.

1.6.1 Volume of Biosolids Pumped (Capacity)

One of the most common positive displacement biosolids pumps is the piston pump. Each stroke of a piston pump "displaces" or pushes out biosolids. Normally, the piston pump is operated at about 50 gpm. In making capacity for positive displacement pump calculations, we use the volume of biosolids pumped equation shown in the following.

$$\text{Vol. of Biosolids Pumped, gal/min} = \left[(0.785)\left(D^2\right)(\text{Stroke Length})\right.$$
$$\left.(7.48 \text{ gal/cu ft}\right][\text{No. of Strokes/min}] \quad (1.17)$$

Example 1.25

Problem:
A biosolids pump has a bore of 6 in and a stroke length of 4 in. If the pump operates at 55 rpm (or strokes per minute), how many gallons per minute are pumped? (Assume 100% efficiency.)

Solution:

$$\text{Vol. of Biosolids Pumped} = (\text{Gallons pumped/stroke})\ (\text{No. of Strokes/minute})$$
$$= \left[(0.785)\left(D^2\right)(\text{Stroke Length})\ (7.48 \text{ gal/cu ft})\right]\ [\text{Strokes/min}]$$
$$= \left[(0.785)\ (0.5 \text{ ft})\ (0.5 \text{ ft})\ (0.33 \text{ ft})\ (7.48 \text{ gal/cu ft})\right]\ [55 \text{ strokes/min}]$$
$$= (0.48 \text{ gal/stroke})\ (50 \text{ strokes/min})$$
$$= 26.6 \text{ gpm}$$

Example 1.26

Problem:
A biosolids pump has a bore of 6 in and a stroke setting of 3 in. The pump operates at 50 rpm. If the pump operates a total of 60 min during a 24 hr period, what is the gallons per day pumping rate? (Assume the piston is 100% efficient.)

First, calculate the gallons per minute pumping rate:

$$\text{Vol. Pumped, gpm} = (\text{Gallons pumped/Stroke})\ (\text{No. of Strokes/minute})$$
$$= \left[(0.785)\ (0.5 \text{ ft})\ (0.5 \text{ ft})\ (0.25 \text{ ft})\right.$$
$$\left.(7.48 \text{ gal/cu ft})\right][50 \text{ Strokes/min}]$$
$$= (0.37 \text{ gal/stroke})\ (50 \text{ strokes/min})$$
$$= 18.5 \text{ gpm}$$

Then convert gallons per minute to gallons per day pumping rate, based on total minutes pumped during 24 hr:

$$= (18.5)\ (60/\text{day})$$
$$= 1110 \text{ gpd}$$

2 Preliminary Treatment Calculations

The initial stage of treatment in the wastewater treatment process (following collection and influent pumping) is *preliminary treatment*. Process selection normally is based upon the expected characteristics of the influent flow. Raw influent entering the treatment plant may contain many kinds of materials (trash), and preliminary treatment protects downstream plant equipment by removing these materials, which could cause clogs, jams, or excessive wear in plant machinery. In addition, the removal of various materials at the beginning of the treatment train saves valuable space within the treatment plant.

Two of the processes used in preliminary treatment include screening and grit removal. However, preliminary treatment may also include other processes, each designed to remove a specific type of material that presents a potential problem for downstream unit treatment processes. These processes include shredding, flow measurement, preaeration, chemical addition, and flow equalization. Except in extreme cases, plant design will not include all these items. In this chapter, we focus on and describe typical calculations used in two of these processes: screening and grit removal.

2.1 WET WELL CAPACITY AND PUMPING RATE

Example 2.1

A wet well is 15 ft long, 8 ft wide, and 12 ft deep. What is the gallon capacity of the wet well?

Solution:

$$\begin{aligned}\text{Volume, gal} &= (\text{l})(\text{w})(\text{d})(7.48\ \text{gal/ft}^3)\\ &= (15\ \text{ft})(8\ \text{ft})(12\ \text{ft})(7.48)\\ &= 10{,}771.2\ \text{gal}\end{aligned}$$

Example 2.2

The maximum capacity of a wet well is 4,850 gal. If the wet well is 10 ft long and 8 ft wide, what is the maximum depth of water in the wet well in feet?

Solution:

$$\begin{aligned}&4850\ \text{gal} = (10\ \text{ft})(8\ \text{ft})(\text{d})(7.48)\\ &4850\ \text{gal} = (\text{d})(598.4)\\ &4850\ \text{gal}/598.4 = 8.1\ \text{ft depth}\end{aligned}$$

DOI: 10.1201/9781003354314-2

Example 2.3

Problem:
A wet well is 12 ft by 12 ft. With no influent to the well, a pump lowers the water level 1.3 ft during a 4 min pumping test. What is the pumping rate in gallons per minute?

Solution:

$$\text{Pump rate, gpm} = \frac{(l)(w)(d)(7.48\ \text{gal/ft}^3)}{\text{time, min}}$$

$$= \frac{(12\ \text{ft})(12\ \text{ft})(1.3\ \text{ft})(7.48)}{4\ \text{min}}$$

$$= 350.1\ \text{gpm}$$

Example 2.4

Problem:
The water level in a well drops 16 in during a 3 min pumping test. If the wet well is 8 ft by 6 ft, what is the pumping rate in gallons per minute?

Solution:

$$\text{Pumping rate, gpm} = \frac{(8\ \text{ft})(6\ \text{ft})(1.33\ \text{ft})(7.48)}{3\ \text{min}}$$

$$= 159.2\ \text{gpm}$$

2.2 SCREENING

Screening removes large solids, such as rags, cans, rocks, branches, leaves, roots, etc., from the flow before the flow moves on to downstream processes. An average of 0.5–121 cu ft/MG is removed.

2.2.1 Screenings Removal Calculations

Wastewater operators responsible for screenings disposal are typically required to keep a record of the amount of screenings removed from the flow. To keep and maintain accurate screening's records, the volume of screenings withdrawn must be determined. Two methods are commonly used to calculate the volume of screenings withdrawn:

$$\text{Screenings Removed, cu ft/day} = \frac{\text{Screenings, cu ft}}{\text{days}} \tag{2.1}$$

$$\text{Screenings Removed, cu ft/MG} = \frac{\text{Screenings, cu ft}}{\text{Flow, MG}} \tag{2.2}$$

Example 2.5

Problem:
A total of 65 gal of screenings are removed from the wastewater flow during a 24 hr period. What is the screenings removal reported as cubic feet per day?

Solution:
First, convert gallon screenings to cubic feet.

$$\frac{65 \text{ gal}}{7.48 \text{ gal/cu ft}} = 8.7 \text{ cu ft screenings}$$

Next, calculate screenings removed as cubic feet per day:

$$\text{Screenings Removed (cu/ft/day)} = \frac{8.7 \text{ cu ft}}{1 \text{ day}} = 8.7 \text{ cu ft/day}$$

Example 2.6

Problem:
For 1 week, a total of 310 gal of screenings were removed from the wastewater screens. What is the average removal in cubic feet per day?

Solution:
First, gallon screenings must be converted to cubic feet screenings:

$$\frac{310 \text{ gal}}{7.48 \text{ gal/cu ft}} = 41.4 \text{ cu ft screenings}$$

Next, the screenings removal calculation is completed:

$$\text{Screenings Removed, cu ft/day} = \frac{310 \text{ gal}}{7.48 \text{ gal/cu ft}} = 5.9 \text{ cu ft/day}$$

2.2.2 Screenings Pit Capacity Calculations

Recall that detention time may be considered the time required for flow to pass through a basin or tank or the time required to fill a basin or tank at a given flow rate. In screenings pit capacity problems, the time required to fill a screenings pit is being calculated. The equation used in screenings pit capacity problems is given here:

$$\text{Screenings Pit Fill Time, days} = \frac{\text{Volume of Pit, cu ft}}{\text{Screenings Removed, cu ft/day}} \tag{2.3}$$

Example 2.7

Problem:
A screening pit has a capacity of 500 cu ft. (The pit is actually larger than 500 cu ft to accommodate soil for covering.) If an average of 3.4 cu ft of screenings are removed daily from the wastewater flow, in how many days will the pit be full? See Figure 2.1.

Solution:

$$\text{Screenings Pit Fill Time, days} = \frac{\text{Volume of Pit, cu ft}}{\text{Screenings Removed, cu ft/day}}$$

$$= \frac{500\text{ cu ft}}{3.4\text{ cu ft/day}}$$

$$= 147.1\text{ days}$$

Example 2.8

Problem:
A plant has been averaging a screenings removal of 2 cu ft/MG. If the average daily flow is 1.8 MGD, how many days will it take to fill the pit with an available capacity of 125 cu ft? See Figure 2.2.

Solution:
The filling rate must first be expressed as cubic feet per day:

$$\frac{(2\text{ cu ft})\ (1.8\text{ MGD})}{\text{MG}} = 3.6\text{ cu ft/day}$$

$$\text{Screenings Pit Fill Time, days} = \frac{125\text{ cu ft}}{3.6\text{ cu ft/day}}$$

$$= 34.7\text{ days}$$

Example 2.9

Problem:
A screening pit has a capacity of 12 cu yds available for screenings. If the plant removes an average of 2.4 cu ft of screenings per day, in how many days will the pit be filled?

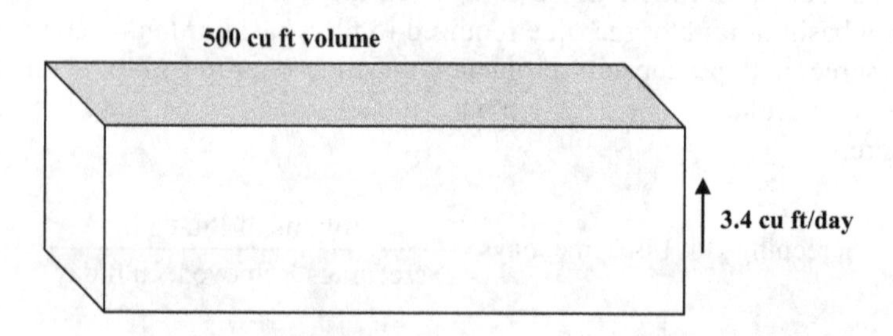

FIGURE 2.1 Screening pit. Refer to example 2.7.

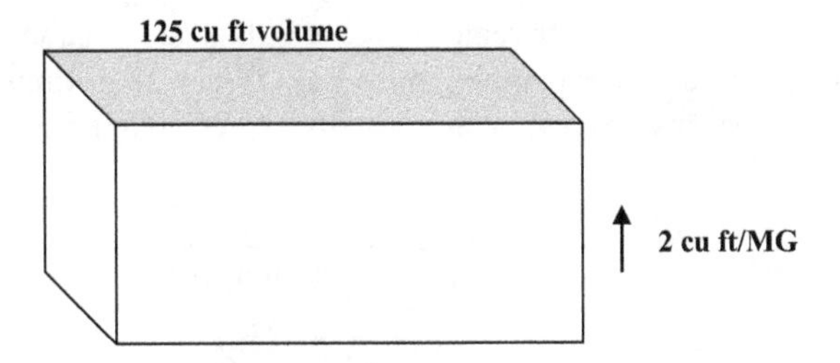

FIGURE 2.2 Screenings pit. Refer to example 2.8.

Solution:
Because the filling rate is expressed as cubic feet per day, the volume must be expressed as cubic foot:

$$(12 \text{ cu yds}) \ (27\text{cu ft/cu yd}) = 324 \text{ cu ft}$$

Now calculate fill time:

$$\text{Screenings Pit Fill Time, days} = \frac{\text{Volume of Pit, cu ft}}{\text{Screenings Removed, cu ft/day}}$$

$$= \frac{324 \text{ cu ft}}{2.4 \text{ cu ft/day}}$$

$$= 135 \text{ days}$$

2.3 GRIT REMOVAL

The purpose of *grit removal* is to remove inorganic solids (sand, gravel, clay, eggshells, coffee grounds, metal filings, seeds, and other similar materials) that could cause excessive mechanical wear. Several processes or devices are used for grit removal, all based on the fact that grit is heavier than the organic solids, which should be kept in suspension for treatment in following unit processes. Grit removal may be accomplished in grit chambers or by the centrifugal separation of biosolids. Processes use gravity/velocity, aeration, or centrifugal force to separate the solids from the wastewater.

2.3.1 Grit Removal Calculations

Wastewater systems typically average 1–15 cu ft of grit per million gallons of flow (sanitary systems: 1 to 4 cu ft/million gal; combined wastewater systems average from 4 to 15 cu ft/million gals of flow), with higher ranges during storm events. Generally, grit is disposed of in sanitary landfills. Because of this process, for planning purposes, operators must keep accurate records of grit removal. Most often, the data are reported as cubic feet of grit removed per million gallons for flow:

$$\text{Grit Removed, cu ft/MG} = \frac{\text{Grit Volume, cu ft}}{\text{Flow, MG}} \tag{2.4}$$

Over a given period, the average grit removal rate at a plant (at least a seasonal average) can be determined and used for planning purposes. Typically, grit removal is calculated as cubic yards, because excavation is normally expressed in terms of cubic yards.

$$\text{Cubic Yards Grit} = \frac{\text{Total Grit, cu ft}}{\text{27 cu ft/cu yd}} \tag{2.5}$$

Example 2.10

Problem:
A treatment plant removes 10 cu ft of grit in 1 day. How many cubic feet of grit are removed per million gallons if the plant flow is 9 MGD?

Solution:

$$\text{Grit Removed, cu ft/MG} = \frac{\text{Grit Volume, cu ft}}{\text{Flow, MG}}$$

$$= \frac{\text{10 cu ft}}{\text{9 MG}} = \text{1.1 cu ft/MG}$$

Example 2.11

Problem:
The total daily grit removed for a plant is 250 gal. If the plant flow is 12.2 MGD, how many cubic feet of grit are removed per MG flow?

Solution:
First, convert gallon grit removed to cubic feet:

$$\frac{\text{250 gal}}{\text{7.48 gal/cu ft}} = \text{33 cu ft}$$

Next, complete the calculation of cubic feet per million gallons:

$$\text{Grit Removed, cu ft/MG} = \frac{\text{Grit Vol., cu ft}}{\text{Flow, MG}}$$

$$= \frac{\text{33 cu ft}}{\text{12.2 MGD}} = \text{2.7 cu ft/MGD}$$

Example 2.12

Problem:
The monthly average grit removal is 2.5 cu ft/MG. If the monthly average flow is 2,500,000 gpd, how many cubic yards must be available for grit disposal if the disposal pit is to have a 90-day capacity?

Solution:
First, calculate the grit generated each day:

$$\left(\frac{\text{2.5 cu ft}}{\text{MG}}\right)(\text{2.5 MGD}) = \text{6.25 cu ft each day}$$

The cubic feet grit generated for 90 days would be:

$$\frac{(6.25 \text{ cu ft})}{\text{day}}(90 \text{ days}) = 562.5 \text{ cu ft}$$

Convert cubic feet to cubic yard grit:

$$\frac{562.5 \text{ cu ft}}{27 \text{ cu ft/cu yd}} = 21 \text{ cu yd}$$

2.3.2 Grit Channel Velocity Calculation

The optimum velocity in sewers is approximately 2 fps at peak flow, because this velocity normally prevents solids from settling from the lines. However, when the flow reaches the grit channel, the velocity should decrease to about 1 fps to permit the heavy inorganic solids to settle. In the example calculations that follow, we describe how the velocity of the flow in a channel can be determined, by the float and stopwatch method and by channel dimensions.

Example 2.13 (Velocity by Float and Stopwatch)

$$\text{Velocity, ft/second} = \frac{\text{Distance Traveled, ft}}{\text{Time required, seconds}} \tag{2.6}$$

Problem:
It takes a float 30 sec to travel 37 ft in a grit channel. What is the velocity of the flow in the channel?

Solution:

$$\text{Velocity, fps} = \frac{37 \text{ ft}}{30 \text{ sec}} = 1.2 \text{ fps}$$

Example 2.14 (Velocity by Flow and Channel Dimensions)

This calculation can be used for a single channel or tank or for multiple channels or tanks with the same dimensions and equal flow. If the flow through each unit of the unit dimensions is unequal, the velocity for each channel or tank must be computed individually.

$$\text{Velocity, fps} = \frac{\text{Flow, MGD} \times 1.55 \text{ cfs/MGD}}{\text{\# Channels in Service} \times \text{Channel Width, ft} \times \text{Water Depth, ft}} \tag{2.7}$$

Problem:
A plant is currently using two girt channels. Each channel is 3 ft wide and has a water depth of 1.3 ft. What is the velocity when the influent flow rate is 4 MGD?

Solution:

$$\text{Velocity, fps} = \frac{4.0\ \text{MGD} \times 1.55\ \text{cfs/MGD}}{2\ \text{Channels} \times 3\ \text{ft} \times 1.3\ \text{ft}}$$

$$\text{Velocity, fps} = \frac{6.2\ \text{cfs}}{7.8\ \text{ft}^2} = 0.79\ \text{fps}$$

✓ **Key Point:** Because 0.79 is within the 0.7–1.4 level, the operator of this unit would not make any adjustments.

✓ **Key Point:** The channel dimensions must always be in feet. Convert inches to feet by dividing by 12 in per foot.

Example 2.15 (Required Settling Time)

This calculation can be used to determine the time required for a particle to travel form the surface of the liquid to the bottom at a given settling velocity. To compute the settling time, settling velocity in feet per second must be provided or determined by experiment in a laboratory.

$$\text{Settling Time, seconds} = \frac{\text{Liquid Depth in ft}}{\text{Settling, Velocity, fps}} \tag{2.8}$$

Problem:

The plant's grit channel is designed to remove sand, which has a settling velocity of 0.080 fps. The channel is currently operating at a depth of 2.3 ft. How many seconds will it take for a sand particle to reach the channel bottom?

Solution:

$$\text{Settling Time, sec} = \frac{2.3\ \text{ft}}{0.080\ \text{fps}} = 28.8\ \text{sec}$$

Example 2.16 (Required Channel Length)

This calculation can be used to determine the length of channel required to remove an object with a specified settling velocity.

$$\text{Required Channel Length} = \frac{\text{Channel Depth, ft} \times \text{Flow Velocity, fps}}{0.080\ \text{fps}} \tag{2.9}$$

Problem:

The plant's grit channel is designed to remove sand, which has a settling velocity of 0.080 fps. The channel is currently operating at a depth of 3 ft. The calculated velocity of flow through the channel is 0.85 fps. The channel is 36 ft long. Is the channel long enough to remove the desired sand particle size?

Solution:

$$\text{Required Channel Length} = \frac{3\text{ ft} \times 0.85\text{ fps}}{0.080\text{ fps}} = 31.9\text{ ft}$$

Yes, the channel is long enough to ensure all the sand will be removed.

2.4 SCREENINGS AND GRIT REMOVAL PRACTICE PROBLEMS

Problem 2.1

A total of 58 gal (58/7.48 = 7.754 cu ft) of screenings are removed from the wastewater flow during a 24 hr period. What is the screenings removal in cubic feet per day?

Solution:

$$\text{Screenings Removed, ft}^3\text{/day} = \frac{\text{Screenings, ft}^3}{\text{Day}}$$

$$= \frac{7.754\text{ ft}^3}{1\text{ day}} = 7.754\text{ ft}^3\text{/d}$$

Problem 2.2

The flow of a treatment plant is 3.8 MGD. If the total of 56 cu ft screenings is removed during a 24 hr period, what is the screenings removal in cubic foot per million gallons?

Solution:

$$\text{Screenings removed, ft}^3\text{/MG} = \frac{\text{screenings, ft}^3\text{/d}}{\text{flow, MGD}}$$

$$= \frac{56\text{ ft}^3}{3.8\text{ MGD}} = 14.74\text{ ft}^3\text{/MG}$$

Problem 2.3

A screening pit has a capacity of 420 cu ft. If an average of 3.5 cu ft of screenings is removed daily from the wastewater flow, in how many days will the pit be full?

Solution:

$$\text{Screening Pit Capacity, days} = \frac{\text{Screening Pit Volume, ft}^3}{\text{Screenings Removed, ft}^3\text{/d}}$$

$$= \frac{420\text{ ft}^3}{3.5\text{ ft}^3} = 120\text{ d}$$

Problem 2.4

A treatment plant averages a screenings removal of 2.2 cu ft/MG. If the average daily flow is 2.5 MGD, how many days will it take to fill a 280 cu ft screening pit?

Solution:

$$2.2\text{ft}^3/\text{MG} = \frac{\text{x}}{2.5\text{ MGD}}$$

$$(2.2)(2.5) = 5.5\text{ ft}^3/\text{d}$$

$$\text{Pit capacity, days} = \frac{280\text{ ft}^3}{5.5\text{ ft}^3/\text{d}}$$
$$= 51\text{ days}$$

Problem 2.5

A treatment plant removes 14 cu ft of grit in 1 day. How many cubic feet of grit are removed if the plant flow is 9 MGD?

Solution:

$$\text{Grit Removal, ft}^3/\text{MG} = \frac{\text{Grit Volume, ft}^3/\text{d}}{\text{flow, MGD}}$$
$$= \frac{14\text{ ft}^3/\text{d}}{9\text{ MGD}} = 1.55\text{ ft}^3/\text{MG}$$

Problem 2.6

The total daily grit removal for a plant is 280 gal. If the flow is 12 MGD, find the grit removal in cubic feet per million gallons.

Solution:

$$280/7.48 = 37.43\text{ ft}^3/\text{d}$$

$$\text{Grit Removal, ft}^3/\text{MG} = \frac{37.43\text{ ft}^3/\text{d}}{12\text{ MGD}} = 3.12\text{ ft}^3/\text{MG}$$

Problem 2.7

A grit channel 36 in (3 ft) wide has water flowing to a depth of 1 ft. If the velocity of the wastewater is 1.3 ft/sec, what is the flow in the channel in cubic feet per second and gallons per minute?

Solution:

$$
\begin{aligned}
Q,\ \text{cfs} &= (\text{w, ft})(\text{d, ft})(\text{vel, fps}) \\
&= (3\ \text{ft})(1\ \text{ft})(1.3\ \text{fps}) \\
&= 3.9\ \text{ft}^3/\text{sec} \\
&= (3.9)(7.48)(60) = 1750.3\ \text{gpm}
\end{aligned}
$$

Problem 2.8

A grit channel is 3 ft wide, 60 ft long, with water flowing to a depth of 18 in (1.5 feet). What is the feet per minute velocity through the channel if the flow is 200 gpm?

Solution:

$$
\begin{aligned}
&200\ \text{gpm}/7.48 = 26.7\ \text{ft}^3/\text{min} \\
&26.7\ \text{ft}^3/\text{min} = (3\ \text{ft})(1.5\ \text{ft})(\text{velocity}) \\
&26.7\ \text{ft}^3/\text{min} = (4.5)(\text{velocity}) \\
&26.7/4.5 = 5.9\ \text{fpm} = \text{Velocity}
\end{aligned}
$$

Problem 2.9

A screenings pit has a capacity of 700 cu ft. If an average of 2.8 cu ft of screenings is removed daily from the wastewater flow, in how many days will the pit be full?

Solution:

$$
\text{days} = \frac{700\ \text{ft}^3}{2.8\ \text{ft}^3/\text{d}} = 250\ \text{days}
$$

Problem 2.10

A treatment plant removes 11 cu ft of grit in a day. If the plant flow is 9 MGD, what is the grit removal expressed in cubic feet per million gallons?

Solution:

$$
\text{Grit removal} = \frac{11\ \text{ft}^3/\text{d}}{9\ \text{MGD}} = 1.22\ \text{ft}^3/\text{MG}
$$

3 Primary Treatment Calculations

Primary treatment (primary sedimentation or clarification) should remove both settleable organic and floatable solids. Poor solids removal during this step of treatment may cause organic overloading of the biological treatment processes following primary treatment. Normally, each primary clarification unit can be expected to remove 90 to 95% of settleable solids, 40 to 60% of the total suspended solids, and 25 to 35% of BOD.

3.1 PROCESS CONTROL CALCULATIONS

As with many other wastewater treatment plant unit processes, several process control calculations may be helpful in evaluating the performance of the primary treatment process. Process control calculations are used in the sedimentation process to determine:

- Percent removal
- Hydraulic detention time
- Surface loading rate (surface settling rate; SLR)
- Surface overflow rate (SOR)
- Weir overflow rate (weir loading rate; WOR)
- Biosolids pumping
- Percent total solids (% TS)
- BOD and SS removed, lb/day

In the following sections, we take a closer look at a few of these process control calculations and example problems.

✓ **Key Point:** The calculations presented in the following sections allow you to determine values for each function performed. Again, keep in mind that an optimally operated primary clarifier should have values in an expected range. Recall that the expected range percent removal for a primary clarifier is:

- Settleable solids 90–95%
- Suspended solids 40–60%
- BOD 25–35%

The expected range of hydraulic detention time for a primary clarifier is 1 to 3 hr. The expected range of surface loading/settling rate for a primary clarifier is 600 to 1,200 gpd/sq ft (ballpark estimate). The expected range of weir overflow rate for a primary clarifier is 10,000 to 20,000 gpd/ft.

DOI: 10.1201/9781003354314-3

Surface Loading Rate (Surface Settling Rate/Surface Overflow Rate)

Surface loading rate is the number of gallons of wastewater passing over 1 sq ft of tank per day (see Figure 3.1). This can be used to compare actual conditions with design. Plant designs generally use a surface loading rate of 300 to 1,200 gal/day/ft².

$$\text{Surface Loading Rate, gpd/ft}^2 = \frac{\text{gal/day}}{\text{Surface Tank Area, ft}^2} \tag{3.1}$$

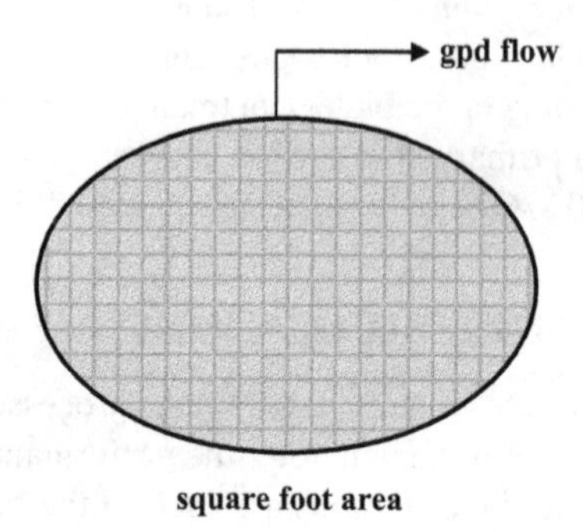

FIGURE 3.1 Primary clarifier.

Example 3.1

Problem:

The circular settling tank has a diameter of 120 ft. If the flow to the unit is 4.5 MGD, what is the surface loading rate in gal/day/ft² (see Figure 3.2)?

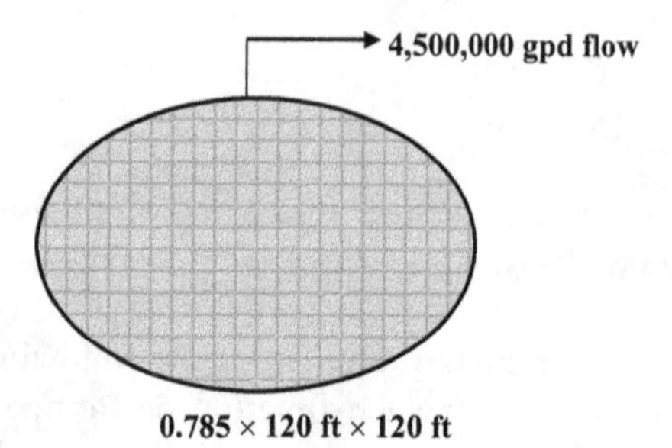

FIGURE 3.2 Refer to example 3.1.

Solution:

$$\text{Surface Loading Rate} = \frac{4.5\text{ MGD} \times 1{,}000{,}000\text{ gal/MGD}}{0.785 \times 120\text{ ft} \times 120\text{ ft}} = 398\text{ gpd/ft}^2$$

Example 3.2

Problem:

A circular clarifier has a diameter of 50 ft. If the primary effluent flow is 2,150,000 gpd, what is the surface overflow rate in gallons per day per square foot?

Solution:

✓ **Key Point:** Remember that area = (0.785) (50 ft) (50 ft)

$$\text{Surface Overflow Rate} = \frac{\text{Flow, gpd}}{\text{Area, sq ft}}$$

$$= \frac{2{,}150{,}000}{(0.785)\ (50\text{ ft})\ (50\text{ ft})} = 1{,}096 \text{ gpd/sq ft}$$

Example 3.3

Problem:

A sedimentation basin 90 ft by 20 ft receives a flow of 1.5 MGD. What is the surface overflow rate in gallons per day per square foot (see Figure 3.3)?

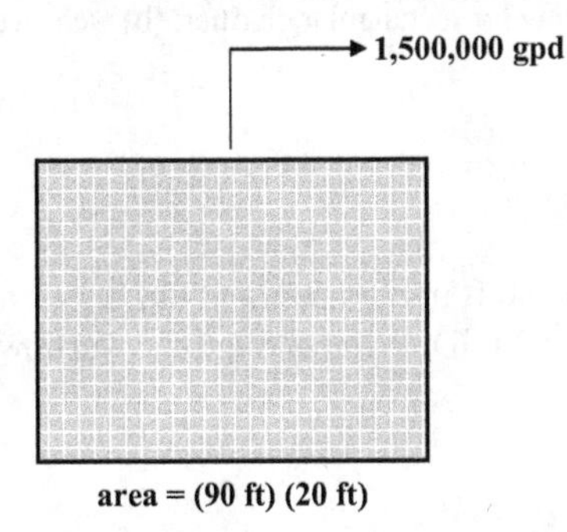

FIGURE 3.3 Refer to example 3.3.

Solution:

$$\text{Surface Overflow Rate} = \frac{\text{Flow, gpd}}{\text{Area, sq ft}}$$

$$= \frac{1{,}500{,}000 \text{ gpd}}{(90\text{ ft})\ (20\text{ ft})}$$

$$= 833 \text{ gpd/sq ft}$$

3.1.1 Weir Overflow Rate (Weir Loading Rate)

A *weir* is a device used to measure wastewater flow (see Figure 3.4). *Weir overflow rate (weir loading rate)* is the amount of water leaving the settling tank per linear foot of water. The result of this calculation can be compared with design. Normally, weir overflow rates of 10,000 to 20,000 gal/day/ft are used in the design of a settling tank.

$$\textit{Weir Overflow Rate, gpd / ft} = \frac{\text{Flow, gal/day}}{\text{Weir Length, ft}} \tag{3.2}$$

✓ **Key Point:** In calculating weir circumference, use total feet of weir = (3.14) (weir diameter, ft).

A.

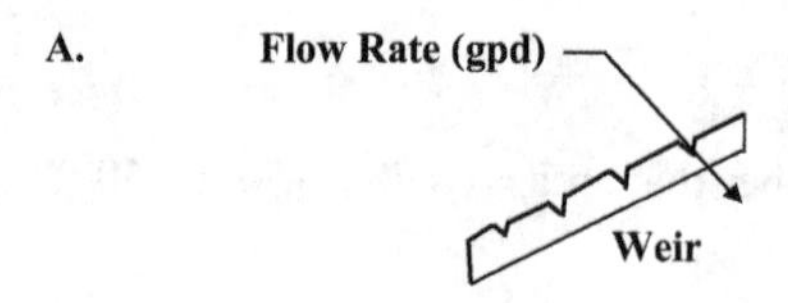

B.

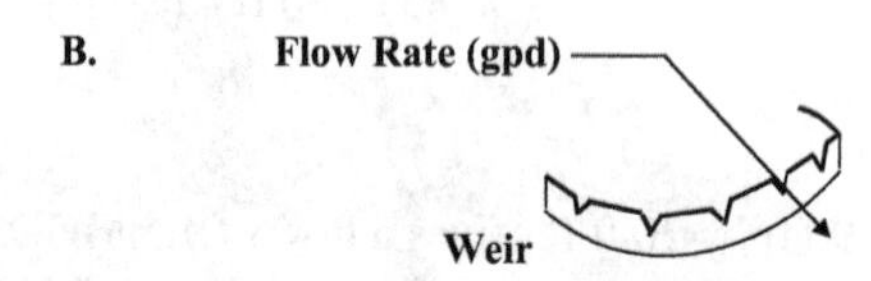

FIGURE 3.4 (a) Weir overflow for rectangular clarifier; (b) weir overflow for circular clarifier.

Example 3.4

Problem:
The circular settling tank is 80 ft in diameter and has a weir along its circumference. The effluent flow rate is 2.75 MGD. What is the weir overflow rate in gallons per day per foot?

Solution:

$$\text{Weir Overflow Rate, gpd/ft} = \frac{2.75 \text{ MGD} \times 1{,}000{,}000 \text{ gal}}{3.14 \times 80 \text{ ft}} = 10{,}947 \text{ gal/day/ft}$$

✓ **Key Point:** Notice that 10,947 gal/day/ft is above the recommended minimum of 10,000.

Example 3.5

Problem:
A rectangular clarifier has a total of 70 ft of weir. What is the weir overflow rate in gallons per day per foot when the flow is 1,055,000 gpd?

Solution:

$$\text{Weir Overflow Rate} = \frac{\text{Flow, gpd}}{\text{Weir Length, ft}}$$

$$= \frac{1{,}055{,}000 \text{ gpd}}{70 \text{ ft}} = 15{,}071 \text{ gpd}$$

3.1.2 Biosolids Pumping

Determination of *biosolids pumping* (the quantity of solids and volatile solids removed from the sedimentation tank) provides accurate information needed for process control of the sedimentation process.

$$\text{Solids Pumped,} = \text{Pump Rate, gpm} \times \text{Pump Time,} \\ \text{min/day} \times 8.34\ \text{lb/gal} \times \%\ \text{Solids} \tag{3.3}$$

$$\text{Volatile Solids/lb/day} = \text{Pump Rate} \times \text{Pump Time} \times 8.34 \\ \times \%\ \text{Solids} \times \%\ \text{Volatile Matter} \tag{3.4}$$

Example 3.6

Problem:
A biosolids pump operates 30 min/hr. The pump delivers 25 gal/min of biosolids. Laboratory tests indicate that the biosolids is 5.3% solids and 68% volatile matter. How many pounds of volatile matter are transferred from the settling tank to the digester? Assume a 24 hr period.

Solution:

$$\text{Pump time} = 30\ \text{min/hr}$$
$$\text{Pump rate} = 25\ \text{gpm}$$
$$\%\ \text{solids} = 5.3\%$$
$$\%\ \text{VM} = 68\%$$

$$\text{Volatile Solids, lb/day} = 25\ \text{gpm} \times (30\ \text{min/hr} \times 24\ \text{hr/day}) \times 8.34\ \text{lb/gal} \times 0.053 \times 0.68$$
$$= 5{,}410\ \text{lb/day}$$

Percent Total Solids (% TS)

Problem:
A settling tank biosolids sample is tested for solids. The sample and dish weigh 73.79 g. The dish alone weighs 21.4 g. After drying, the dish with dry solids weighs 22.4 g. What is the percent total solids (% TS) of the sample?

Solution:

Sample + Dish	73.79 g	Dish + Dry Solids	22.4 g
Dish alone	– 21.4 g	Dish Alone	– 21.4 g
	52.39 g		1.0 g

$$\frac{1.0\ \text{g}}{52.39\ \text{g}} \times 100\% = 1.9\%$$

3.1.3 BOD and SS Removed, Lb/Day

To calculate the pounds of BOD or suspended solids removed each day, we need to know the milligrams per liter BOD or SS removed and the plant flow. Then, we can use the milligrams per liter to pounds per day equation.

$$\text{SS Removed} = \text{mg/L} \times \text{MGD} \times 8.34 \text{ lb/gal} \tag{3.5}$$

Example 3.7

Problem:
If 120 mg/L suspended solids are removed by a primary clarifier, how many pounds per day suspended solids are removed when the flow is 6,250,000 gpd?

Solution:

$$\text{SS Removed} = 120 \text{ mg/L} \times 6.25 \text{ MGD} \times 8.34 \text{ lb/gal} = 6{,}255 \text{ lb/day}$$

Example 3.8

Problem:
The flow to a secondary clarifier is 1.6 MGD. If the influent BOD concentration is 200 mg/L and the effluent BOD concentration is 70 mg/L, how many pounds of BOD are removed daily?

$$\text{lb/day BOD removed} = 200 \text{ mg/L} - 70 \text{ mg/L} = 130 \text{ mg/L}$$

After calculating milligrams per liter BOD removed, calculate pounds per day BOD removed.

$$\text{BOD removed, lb/day} = (130 \text{ mg/L})\ (1.6 \text{ MGD})\ (8.34 \text{ lb/gal}) = 1{,}735 \text{ lb/day}$$

3.2 PRIMARY TREATMENT PRACTICE PROBLEMS

Problem 3.1

The flow to a circular clarifier is 4,000,000 gpd. If the clarifier is 80 ft in diameter and 12 ft deep, what is the clarifier detention time in hours?

Solution:

$$\text{Detention Time, hrs} = \frac{(\text{Vol, gal})(24 \text{ hr/day})}{\text{Flow, gpd}}$$

$$\text{Volume} = (0.785)(80)^2\,(12 \text{ ft})(7.48) = 450{,}954.2 \text{ gal}$$

$$= \frac{(450{,}954.2 \text{ gal})(24)}{4{,}000{,}000 \text{ gpd}} = 2.7 \text{ hrs}$$

Problem 3.2

A circular clarifier has a diameter of 60 ft. If the primary clarifier influent flow is 2,300,000 gpd, what is the surface overflow rate (SOR) in gallons per day per square feet?

Solution:

$$\text{SOR, gpd/ft}^2 = \frac{\text{flow, gpd}}{\text{area, ft}^2}$$

$$= \frac{2{,}300{,}000 \text{ gpd}}{(0.785)(60 \text{ ft})^2} = 813.9 \text{ gpd/ft}^2$$

Problem 3.3

A rectangular clarifier has a total of 200 ft of weir. What is the weir overflow rate (WOR) in gallons per day per foot when the flow 3,730,000 gpd?

Solution:

$$\text{WOR, gpd/ft} = \frac{\text{flow, gpd}}{\text{weir length, ft}}$$

$$= \frac{3{,}730{,}000 \text{ gpd}}{200 \text{ ft}}$$

$$= 18{,}650 \text{ gpd/ft}$$

Problem 3.4

A secondary clarifier, 60 ft in diameter, receives a primary effluent flow of 1,888,000 gpd and a return sludge flow of 530,000 gpd. If the MLSS concentration is 2,700 mg/L, what is the solids loading rate (SLR) in pounds per day per square feet on the clarifier?

Solution:

$$\text{SLR, lb/day/ft}^2 = \frac{(\text{MLSS, mg/L})(\text{Primary Eff.} + \text{RAS flow, MGD})(8.34)}{(0.785)(\text{D, ft})^2}$$

$$= \frac{(2700 \text{ mg/L})(1.888 + 0.530 \text{ MGD})(8.34)}{(0.785)(60\text{-ft})}$$

$$= \frac{54448.524 \text{ lbs/day}}{2826 \text{ ft}^2} = 19.3 \text{ lb/day/ft}^2$$

Problem 3.5

A circular primary clarifier has a diameter of 55 ft. If the influent flow to the clarifier is 2.50 MGD, what is the surface overflow rate in gallons per day per square feet?

Solution:

$$\text{SOR, gpd/ft}^2 = \frac{2{,}500{,}000 \text{ gpd}}{(0.785)(55 \text{ ft})^2} = 1053 \text{ gpd/ft}^2$$

4 Trickling Filter Calculations

The *trickling filter process* (see Figure 4.1) is one of the oldest forms of dependable biological treatment for wastewater. By its very nature, the trickling filter has its advantages over other unit processes. For example, it is a very economical and dependable process for treatment of wastewater prior to discharge. Capable of withstanding periodic shock loading, process energy demands are low because aeration is a natural process. As shown in Figure 4.2, trickling filter operation involves spraying wastewater over a solid media such as rock, plastic, or redwood slats (or laths). As the wastewater trickles over the surface of the media, a growth of microorganisms (bacteria, protozoa, fungi, algae, helminths or worms, and larvae) develops. This growth is visible as a shiny slime very similar to the slime found on rocks in a stream. As wastewater passes over this slime, the slime adsorbs the organic (food) matter. This organic matter is used for food by the microorganisms. At the same time, air moving through the open spaces in the filter transfers oxygen to the wastewater. This oxygen is then transferred to the slime to keep the outer layer aerobic. As the microorganisms use the food and oxygen, they produce more organisms, carbon dioxide, sulfates, nitrates, and other stable by-products; these materials are then discarded from the slime back into the wastewater flow and are carried out of the filter.

4.1 TRICKLING FILTER PROCESS CALCULATIONS

Several calculations are useful in the operation of trickling filters: these include hydraulic loading, organic loading, and biochemical oxygen demand (BOD) and suspended solids (SS) removal. Each type of trickling filter is designed to operate with specific loading levels. These levels vary greatly depending on the filter classification. To operate the filter properly, filter loading must be within the specified levels. The main three loading parameters for the trickling filter are hydraulic loading, organic loading, and recirculation ratio.

4.1.1 Hydraulic Loading Rate

Calculating the *hydraulic loading rate* is important in accounting for both the primary effluent as well as the recirculated trickling filter effluent. These are combined before being applied to the filter surface. The hydraulic loading rate is calculated based on filter surface area. The normal hydraulic loading rate ranges for standard rate and high-rate trickling filters are:

- Standard rate: 25–100 gpd/sq ft, or 1–40 MGD/acre
- High rate: 100–1,000 gpd/sq ft, or 4–40 MGD/acre

DOI: 10.1201/9781003354314-4

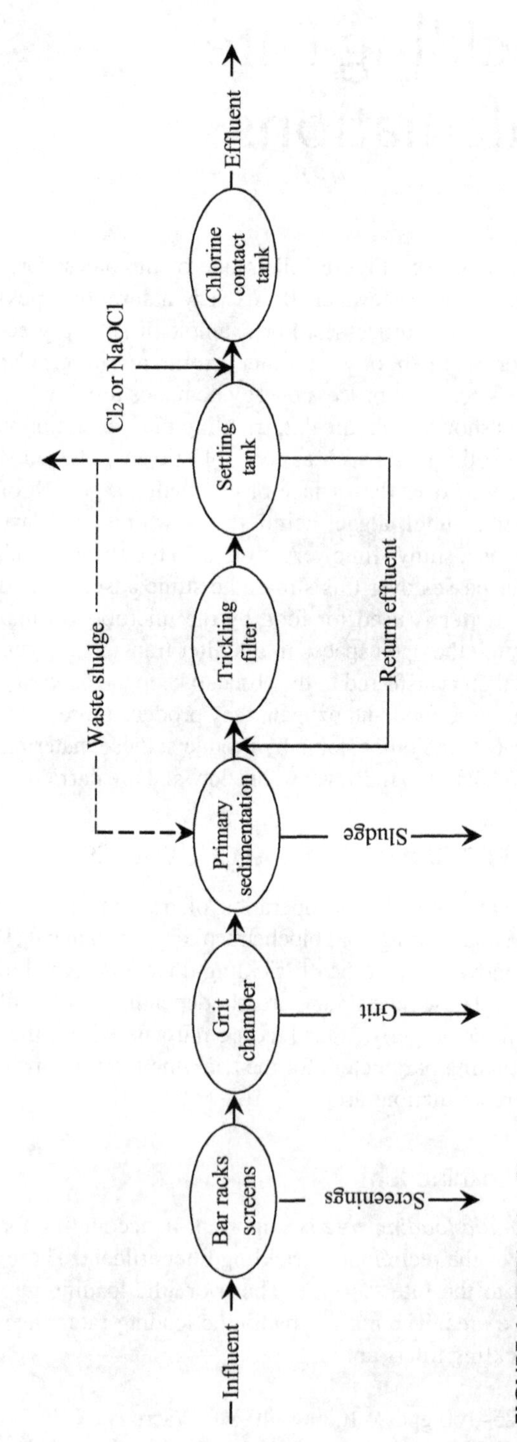

FIGURE 4.1 Trickling filter system.

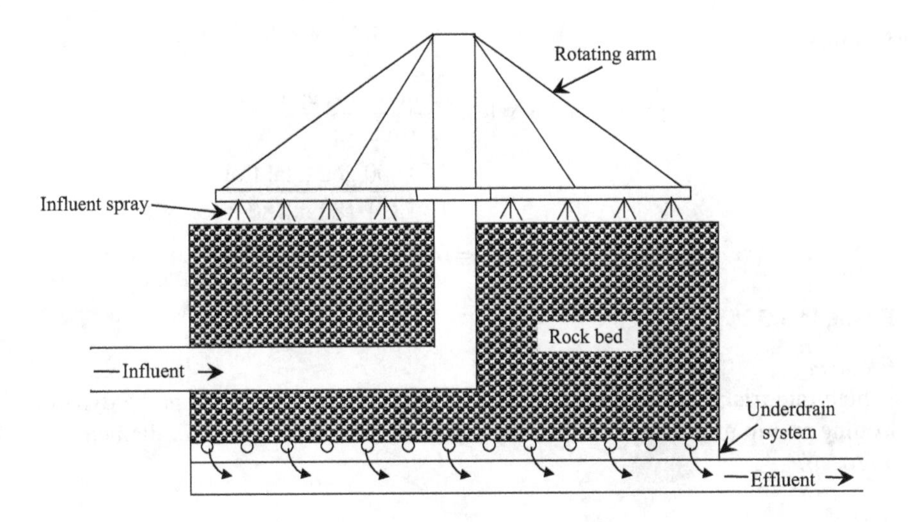

FIGURE 4.2 Cross section of a trickling filter.

✓ **Key Point:** If the hydraulic loading rate for a particular trickling filter is too low, septic conditions will begin to develop.

Example 4.1

Problem:
A trickling filter 80 ft in diameter is operated with a primary effluent of 0.588 MGD and a recirculated effluent flow rate of 0.660 MGD. Calculate the hydraulic loading rate on the filter in unit gallons per day per square feet.

Solution:
The primary effluent and recirculated trickling filter effluent are applied together across the surface of the filter; therefore, 0.588 MGD + 0.660 MGD = 1.248 MGD = 1,248,000 gpd.

$$\begin{aligned}\text{Circular surface area} &= 0.785 \times (\text{diameter})^2 \\ &= 0.785 \times (80\text{ ft})^2 \\ &= 5{,}024\text{ ft}^2\end{aligned}$$

$$\frac{1{,}248{,}000\text{ gpd}}{5{,}024\text{ ft}^2} = 248.4\text{ gpd/ft}^2$$

Example 4.2

Problem:
A trickling filter 80 ft in diameter treats a primary effluent flow of 550,000 gpd. If the recirculated flow to the clarifier is 0.2 MGD, what is the hydraulic loading on the trickling filter?

Solution:

$$\text{Hydraulic loading rate} = \frac{\text{Total Flow, gpd}}{\text{Area, sq ft}}$$

$$= \frac{750{,}000 \text{ gpd total flow}}{(0.785)\ (80 \text{ ft})(80 \text{ ft})}$$

$$= 149 \text{ gpd/ sq ft}$$

Example 4.3

Problem:
A high-rate trickling filter receives a daily flow of 1.8 MGD. What is the dynamic loading rate in million gallons per day per acre if the filter is 90 ft in diameter and 5 ft deep?

Solution:

$$(0.785)\ (90 \text{ ft})\ (90 \text{ ft}) = 6{,}359 \text{ sq ft}$$

$$\frac{6{,}359 \text{ sq ft}}{43{,}560 \text{ sq ft/ac}} = 0.146 \text{ acres}$$

$$\text{Hydraulic Loading Rate} = \frac{1.8 \text{ MGD}}{0.146 \text{ acres}} = 12.3 \text{ MGD/ac}$$

✓ **Key Point:** When hydraulic loading rate is expressed as million gallons per day per acre, this is still an expression of gallon flow over surface area of trickling filter.

4.1.2 Organic Loading Rate

Trickling filters are sometimes classified by the *organic loading rate* applied. The organic loading rate is expressed as a certain amount of BOD applied to a certain volume of media. In other words, the organic loading is defined as the pounds of BOD_5 or chemical oxygen demand (COD) applied per day per 1,000 cu ft of media—a measure of the amount of food being applied to the filter slime. To calculate the organic loading on the trickling filter, two things must be known: the pounds of BOD or COD being applied to the filter media per day and the volume of the filter media in 1,000 cu ft units. The BOD and COD contribution of the recirculated flow is not included in the organic loading.

Example 4.4

Problem:
A trickling filter, 60 ft in diameter, receives a primary effluent flow rate of 0.440 MGD. Calculate the organic loading rate in units of pounds of BOD applied per day per 1,000 cu ft of media volume. The primary effluent BOD concentration is 80 mg/L. The media depth is 9 ft.

Solution:

$$0.440\ \text{MGD} \times 80\ \text{mg/L} \times 8.34\ \text{lb/gal} = 293.6\ \text{lb of BOD applied/d}$$

$$\text{Surface Area} = 0.785 \times (60)^2 = 2{,}826\ \text{ft}^2$$

$$\text{Area} \times \text{Depth} \times \text{Volume}$$

$$2{,}826\ \text{ft}^2 \times 9\ \text{ft} = 25{,}434\ \text{cu ft (TF Volume)}$$

✓ **Key Point:** To determine the pounds of BOD per 1,000 cu ft in a volume of thousands of cubic feet, we must set up the equation as shown in the following.

$$\frac{293.6\ \text{lb BOD/d}}{25{,}434\ \text{ft}^3} \times 1000$$

Regrouping the numbers and the units together:

$$\frac{293.6\ \text{lb BOD/d} \times 1{,}000}{25{,}434\ \text{ft}^3} \times \frac{\text{lb BOD/d}}{1{,}000\ \text{ft}^3} = 11.5 \frac{\text{lb BOD/d}}{1{,}000\ \text{ft}^3}$$

4.1.3 BOD and SS Removed

To calculate the pounds of BOD or suspended solids removed each day, we need to know the milligrams per liter BOD and SS removed and the plant flow.

Example 4.5

Problem:
If 120 mg/L suspended solids are removed by a trickling filter, how many pounds per day suspended solids are removed when the flow is 4 MGD?

Solution:

$$(\text{mg/L})\ (\text{MGD flow}) \times 8.34\ \text{lb/gal}) = \text{lb/day}$$

$$(120\ \text{mg/L})\ (4.0\ \text{MGD}) \times (8.34\ \text{lb/gal}) = 4{,}003\ \text{lb SS/day}$$

Example 4.6

Problem:
The 3,500,000 gpd influent flow to a trickling filter has a BOD content of 185 mg/L. If the trickling filter effluent has a BOD content of 66 mg/L, how many pounds of BOD are removed daily?

Solution:

$$(\text{mg/L})\ (\text{MGD flow})\ (8.34\ \text{lb/gal}) = \text{lb/day removed}$$

$$185\ \text{mg/L} - 66\ \text{mg/L} = 119\ \text{mg/L}$$

$$(119\ \text{mg/L})\ (3.5\ \text{MGD})\ (8.34\ \text{lb/gal}) = 3{,}474\ \text{lb/day removed}$$

4.1.4 Recirculation Flow

Recirculation in trickling filters involves the return of filter effluent back to the head of the trickling filter. It can level flow variations and assist in solving operational problems, such as ponding, filter flies, and odors. The operator must check the rate of recirculation to ensure that it is within design specifications. Rates above design specifications indicate hydraulic overloading; rates under design specifications indicate hydraulic underloading. The *trickling filter recirculation ratio* is the ratio of the recirculated trickling filter flow to the primary effluent flow. The trickling filter recirculation ratio may range from 0.5:1 (.5) to 5:1 (5). However, the ratio is often found to be 1:1 or 2:1.

$$\text{Recirculation} = \frac{\text{Recirculated Flow, MGD}}{\text{Primary Effluent Flow, MGD}} \tag{4.1}$$

Example 4.7

Problem:
A treatment plant receives a flow of 3.2 MGD. If the trickling filter effluent is recirculated at the rate of 4.50 MGD, what is the recirculation ratio?

Solution:

$$\begin{aligned}\text{Recirculation Ratio} &= \frac{\text{Recirculated Flow, MGD}}{\text{Primary Effluent Flow, MGD}}\\ &= \frac{4.5\text{ MGD}}{3.2\text{ MGD}}\\ &= 1.4\text{ Recirculation Ratio}\end{aligned}$$

Example 4.8

Problem:
A trickling filter receives a primary effluent flow of 5 MGD. If the recirculated flow is 4.6 MGD, what is the recirculation ratio?

Solution:

$$\begin{aligned}\text{Recirculation Ratio} &= \frac{\text{Recirculated Flow, MGD}}{\text{Primary Effluent Flow, MGD}}\\ &= \frac{4.6\text{ MGD}}{5\text{ MGD}}\\ &= 0.92\text{ Recirculation Ratio}\end{aligned}$$

4.2 TRICKLING FILTER PRACTICE PROBLEMS

Problem 4.1

A trickling filter, 90 ft in diameter, treats a primary effluent flow of 530,000 gpd. If the recirculated flow to the trickling filter is 110,000 gpd, what is the hydraulic loading rate (HLR) on the filter in gallons per day per square foot?

Solution:

$$\text{HLR, gpd/ft}^2 = \frac{\text{Primary effluent flow} + \text{Recirculation flow, gpd}}{\text{Area, ft}^2}$$

$$= \frac{530{,}000 + 110{,}000 \text{ gpd}}{(0.785)(90)^2} = 100.7 \text{ gpd/ft}^2$$

Problem 4.2

A trickling filter, 70 ft in diameter, treats a primary effluent flow of 630,000 gpd. If the recirculated flow to the trickling filter is 100,000 gpd, what is the hydraulic loading rate (HLR) in gallons per day per square feet on the trickling filter?

Solution:

$$\text{HLR, gpd/ft}^2 = \frac{630{,}000 + 100{,}000 \text{ gpd}}{(0.785)(70 \text{ ft})^2} = 189.8 \text{ gpd/ft}^2$$

Problem 4.3

A trickling filter, 80 ft in diameter, with a media depth of 5 ft, receives a flow of 1,100,000 gpd. If the BOD concentration of the primary effluent is 160 mg/L, what is the organic loading rate (OLR) on the trickling filter in pound BOD per day per 1,000 cu ft.

Solution:

$$\text{OLR, lb/d/1000 ft}^3 = \frac{(\text{BOD, mg/L})(\text{Q, MGD})(8.34)}{\dfrac{(0.785)(80 \text{ ft})^2(\text{d, ft})}{1000}}$$

$$= \frac{(160 \text{ mg/L})(1.1 \text{ MGD})(8.34)}{\dfrac{(0.785)(80 \text{ ft})^2(5 \text{ ft})}{1000}}$$

$$= 58.4 \text{ lb/d/1000 ft}^3$$

5 Rotating Biological Contactors (RBCs)

The *rotating biological contactor (RBC)* is a variation of the attached growth idea provided by the trickling filter (see Figure 5.1). Still relying on microorganisms that grow on the surface of a medium, the RBC is instead a *fixed film* biological treatment device. The basic biological process, however, is similar to that occurring in trickling filters. An RBC consists of a series of closely spaced, mounted side by side circular plastic synthetic disks, typically about 11.5 ft in diameter (see Figure 5.2). Attached to a rotating horizontal shaft, approximately 40% of each disk is submersed in a tank that contains the wastewater to be treated. As the RBC rotates, the attached biomass film (zoogleal slime) that grows on the surface of the discs moves into and out of the wastewater. While submerged in the wastewater, the microorganisms absorb organics; while they are rotated out of the wastewater, they are supplied with needed oxygen for aerobic decomposition. As the zoogleal slime re-enters the wastewater, excess solids and waste products are stripped off the media as *sloughings*. These sloughings are transported with the wastewater flow to a settling tank for removal.

5.1 RBC PROCESS CONTROL CALCULATIONS

Several process control calculations may be useful in the operation of an RBC. These include soluble BOD, total media area, organic loading rate, and hydraulic loading. Settling tank calculations and biosolids pumping calculations may be helpful for evaluation and control of the settling tank following the RBC.

5.1.1 Hydraulic Loading Rate

The manufacturer normally specifies the RBC media surface area, and the hydraulic loading rate is based on the media surface area, usually in square feet (ft^2). Hydraulic loading is expressed in terms of gallons of flow per day per square foot of media. This calculation can be helpful in evaluating the current operating status of the RBC. Comparison with design specifications can determine if the unit is hydraulically over- or underloaded. Hydraulic loading on an RBC can range from 1 to 3 gpd/ft^2.

DOI: 10.1201/9781003354314-5

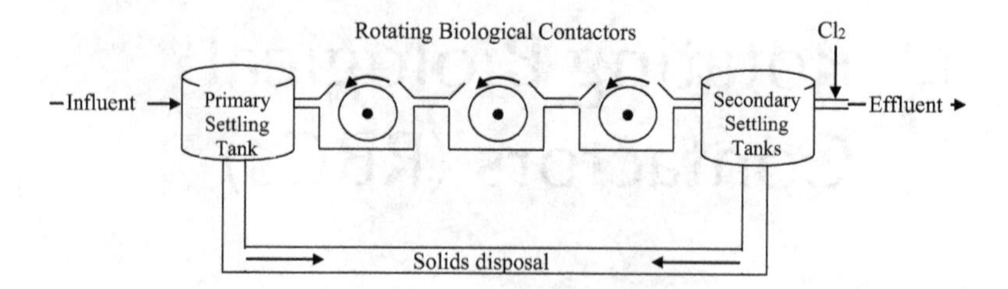

FIGURE 5.1 Rotating biological contactor (RBC) treatment system.

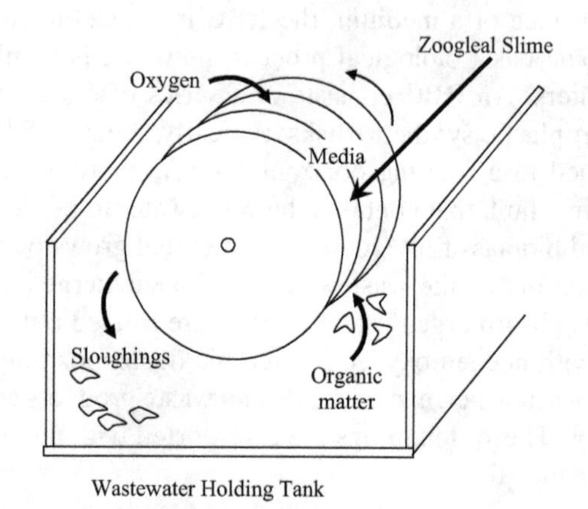

FIGURE 5.2 Rotating Biological Contactor (RBC).

Example 5.1

Problem:

An RBC treats a primary effluent flow rate of 0.244 MGD. What is the hydraulic loading rate in gallons per day per square foot if the media surface area is 92,600 ft²?

Solution:

$$\frac{244{,}000 \text{ gpd}}{92{,}600 \text{ ft}^2} = 2.63 \text{ gpd/ft}^2$$

Example 5.2

Problem:

An RBC treats a flow of 3.5 MGD. The manufacturer's data indicate a media surface area of 750,000 sq ft. What is the hydraulic loading rate on the RBC?

Solution:

$$\text{Hydraulic Loading Rate} = \frac{\text{Flow, gpd}}{\text{Media Area, sq ft}}$$

$$= \frac{3{,}500{,}000 \text{ gpd}}{750{,}000 \text{ sq ft}} = 4.7 \text{ sq ft}$$

Example 5.3

Problem:
A rotating biological contactor treats a primary effluent flow of 1,350,000 gpd. The manufacturer's data indicates that the media surface area is 600,000 sq ft. What is the hydraulic loading rate on the filter?

$$\text{Hydraulic Loading Rate} = \frac{\text{Flow, gpd}}{\text{Area, sq ft}}$$

$$= \frac{3{,}500{,}000 \text{ gpd}}{600{,}000 \text{ sq ft}} = 2.3 \text{ sq ft}$$

5.1.2 Soluble BOD

The *soluble BOD* concentration of the RBC influent can be determined experimentally in the laboratory, or it can be estimated using the suspended solids concentration and the "K" factor. The "K" factor is used to approximate the BOD (particulate BOD) contributed by the suspended matter. The *K* factor must be provided or determined experimentally in the laboratory. The *K* factor for domestic wastes is normally in the range of 0.5 to 0.7.

$$\text{Soluble BOD}_5 = \text{Total BOD}_5 - \left(K\,\text{Factor} \times \text{Total Suspended Solids}\right) \tag{5.1}$$

Example 5.4

Problem:
The suspended solids concentration of a wastewater is 250 mg/L. If the amount of *K*-value at the plant is 0.6, what is the estimated particulate biochemical oxygen demand (BOD) concentration of the wastewater?

Solution:

✓ **Key Point:** The K-value of 0.6 indicates that about 60% of the suspended solids are organic suspended solids (particulate BOD).

$$(250 \text{ mg/L})\ (0.6) = 150 \text{ mg/L Particulate BOD}$$

Example 5.5

Problem:
A rotating biological contactor receives a flow of 2.2 MGD with a BOD content of 170 mg/L and suspended solids (SS) concentration of 140 mg/L. If the K-value is 0.7, how many pounds of soluble BOD enter the RBC daily?

Solution:

$$\text{Total BOD} = \text{Particulate BOD} + \text{Soluble BOD}$$
$$170\ \text{mg/L} = (140\ \text{mg/L})\ (0.7) + \text{xmg/L}$$
$$170\ \text{mg/L} = 98\ \text{mg/L} + \text{xmg/}$$
$$170\ \text{mg/L} - 98\ \text{mg/L} = \text{x}$$
$$\text{x} = 72\ \text{mg/L Soluble BOD}$$

Now, pounds per day soluble BOD may be determined:

$$(\text{mg/L Soluble BOD})\ (\text{MGD Flow})\ (8.34\ \text{lb/gal}) = \text{lb/day}$$
$$(72\ \text{mg/L})\ (2.2\ \text{MGD})\ (8.34\ \text{lb/gal}) = 1{,}321\ \text{lb/day soluble BOD}$$

Example 5.6

Problem:
The wastewater entering a rotating biological contactor has a BOD content of 210 mg/L. The suspended solids content is 240 mg/L. If the K-value is 0.5, what is the estimated soluble BOD (mg/L) of the wastewater?

$$\text{Total BOD, mg/L} = \text{Particulate BOD, mg/L} + \text{Soluble BOD, mg/L}$$
$$210\ \text{mg/L} = (240\ \text{mg/L})\ (0.5) + \text{xmg/L}$$

BOD SS Soluble BOD

$$210\ \text{mg/L} = 120\ \text{mg/L} + \text{xmg/L}$$

Soluble BOD

$$210 - 120 = \text{x}$$
$$\text{Soluble BOD } 90\ \text{mg/L} = \text{x}$$

5.1.3 Organic Loading Rate

The *organic loading rate* can be expressed as total BOD loading in pounds per day per 1,000 sq ft of media. The actual values can then be compared with plant design specifications to determine the current operating condition of the system.

$$\text{Organic Loading Rate} = \frac{\text{Sol. BOD} \times \text{Flow, MGD} \times 8.34\ \text{lb/gal} \times 1000}{\text{Media Area, 1,000 sq ft}} \qquad (5.2)$$

Example 5.7

Problem:
A rotating biological contactor (RBC) has a media surface area of 500,000 sq ft and receives a flow of 1,000,000 gpd. If the soluble BOD concentration of the primary effluent is 160 mg/L, what is the organic loading on the RBC in pounds per day per 1,000 sq ft?

Solution:

$$\text{Organic Loading Rate} = \frac{\text{Sol. BOD, lb/day} \times 1000}{\text{Media Area, 1,000 sq ft}}$$

$$= \frac{(160\text{ mg/L})(1.0\text{ MGD})\ (8.34\text{ lb/gal}) \times 1000}{500 \times 1{,}000\text{ sq ft}}$$

$$= \frac{2.7\text{ lb/day Sol. BOD}}{1{,}000\text{ sq ft}}$$

Example 5.8

Problem:
The wastewater flow to an RBC is 3,000,000 gpd. The wastewater has a soluble BOD concentration of 120 mg/L. The RBC consists of six shafts (each 110,000 sq ft), with two shafts comprising the first stage of the system. What is the organic loading rate in pounds per day per 1,000 sq ft on the first stage of the system?

Solution:

$$\text{Organic Loading Rate} = \frac{\text{Sol. BOD, lb/day} \times 1000}{\text{Media Area, 1,000 sq ft}}$$

$$= \frac{(120\text{ mg/L})(3.0\text{ MGD})\ (8.34\text{ lb/gal}) \times 1000}{220 \qquad 1{,}000\text{ sq ft}}$$

$$= 13.6\text{ lb Sol. BOD/day/1,000 sq ft}$$

5.1.4 Total Media Area

Several process control calculations for the RBC use the total surface area of all the stages within the train. As was the case with the soluble BOD calculation, plant design information or information supplied by the unit manufacturer must provide the individual stage areas (or the total train area), because physical determination of this would be extremely difficult.

$$\text{Total Area} = 1^{\text{st}}\text{stage Area} + 2^{\text{nd}}\text{Stage Area} + \ldots + n\text{th Stage Area} \tag{5.3}$$

5.2 RBC PRACTICE PROBLEMS

Problem 5.1

An RBC treats a primary effluent flow rate of 0.240 MGD. What is the hydraulic loading rate in gallons per day per square foot if the media surface area is 92,600 ft^2?

Solution:

$$\frac{240{,}000 \text{ gpd}}{92{,}000 \text{ ft}^2} = 2.61 \text{ gpd/ft}^2$$

Problem 5.2

An RBC treats a flow of 3.5 MGD. The manufacturer's data indicate a media surface area of 710,000 ft^2. What is the hydraulic loading rate on the RBC?

Solution:

$$\text{Hydraulic Loading Rate} = \frac{\text{Flow, gpd}}{\text{Media Area, ft}^2}$$

$$= \frac{3{,}500{,}000 \text{ gpd}}{710{,}000 \text{ ft}^2} = 4.9 \text{ ft}^2$$

Problem 5.3

A rotating biological contactor treats a primary effluent flow of 1,300,000 gpd. The manufacturer's data indicates that the media surface area is 610,000 ft^2. What is the hydraulic loading rate on the filter?

Solution:

$$\text{Hydraulic Loading Rate} = \frac{\text{Flow, gpd}}{\text{Area, ft}^2}$$

$$= \frac{3{,}500{,}000 \text{ gpd}}{610{,}000 \text{ ft}^2} = 2.13 \text{ ft}^2$$

Problem 5.4

The suspended solids concentration of a wastewater is 250 mg/L. If the amount of *K*-value at the plant is 0.5, what is the estimated particulate biochemical oxygen demand (BOD) concentration of the wastewater?

Solution:

$$(250 \text{ mg/L})\ (0.5) = 125 \text{ mg/L Particulate BOD}$$

Problem 5.5

A rotating biological contactor receives a flow of 2.0 MGD with a BOD content of 170 mg/L and suspended solids (SS) concentration of 120 mg/L. If the K-value is 0.7, how many pounds of soluble BOD enter the RBC daily?

Solution:

$$\text{Total BOD} = \text{Particulate BOD} + \text{Soluble BOD}$$

$$170\text{ mg/L} = (120\text{ mg/L})\ (0.7) + \text{xmg/L}$$

$$170\text{ mg/L} = 84\text{ mg/L} + \text{xmg/}$$

$$170\text{ mg/L} - 84\text{ mg/L} = \text{x}$$

$$\text{x} = 86\text{ mg/L Soluble BOD}$$

Now, lb/day soluble BOD may be determined:

$$(\text{mg/L})\ (\text{Soluble BOD})\ (\text{MGD Flow})\ (8.34\text{ lb/gal}) = \text{lb/day}$$

$$86\text{ mg/L} \times 2.0\text{ MGD} \times 8.34\text{ lb/gal} = 1{,}434.5\text{ lb/day}$$

Problem 5.6

A rotating biological contactor (RBC) has a media surface area of 500,000 ft^2 and receives a flow of 1,000,000 gpd. If the soluble BOD concentration of the primary effluent is 150 mg/L, what is the organic loading on the RBC in pounds per day per 1,000 ft^2?

Solution:

$$\text{Organic Loading Rate} = \frac{\text{Sol. BOD, lb/day} \times 1000}{\text{Media Area, 1,000 ft}}$$

$$= \frac{(150\text{ mg/L})(1.0\text{ MGD})\ (8.34\text{ lb/gal}) \times 1000}{500 \times 1{,}000\text{ ft}^2}$$

$$= \frac{2.5\text{ lb/day Sol. BOD}}{1{,}000\text{ ft}^2}$$

6 Activated Biosolids

The *activated biosolids process* is a man-made process that mimics the natural self-purification process that takes place in steams. In essence, we can state that the activated biosolids treatment process is a "stream in a container." In wastewater treatment, activated biosolids processes are used for both secondary treatment and complete aerobic treatment without primary sedimentation. Activated biosolids refers to biological treatment systems that use a suspended growth of organisms to remove BOD and suspended solids.

The basic components of an activated biosolids sewage treatment system include an aeration tank and a secondary basin, settling basin, or clarifier (see Figure 6.1). Primary effluent is mixed with settled solids recycled form the secondary clarifier and is then introduced into the aeration tank. Compressed air is injected continuously into the mixture through porous diffuses located at the bottom of the tank, usually along one side.

Wastewater is fed continuously into an aerated tank, where the microorganisms metabolize and biologically flocculate the organics. Microorganisms (activated biosolids) are settled from the aerated mixed liquor under quiescent conditions in the final clarifier and are returned to the aeration tank. Left uncontrolled, the number of organisms would eventually become too great; therefore, some must periodically be removed (wasted). A portion of the concentrated solids from the bottom of the settling tank must be removed from the process (waste activated sludge, or WAS). Clear supernatant from the final settling tank is the plant effluent.

6.1 ACTIVATED BIOSOLIDS PROCESS CONTROL CALCULATIONS

As with other wastewater treatment unit processes, process control calculations are important tools used by the operator to control and optimize process operations. In this chapter, we review many of the most frequently used activated biosolids process calculations.

6.1.1 Moving Averages

When performing process control calculations, the use of a seven-day *moving average* is recommended. The moving average is a mathematical method to level the impact of any one test result. The moving average is determined by adding all the test results collected during the past 7 days and dividing by the number of tests.

$$\text{Moving Average} = \frac{\text{Test 1} + \text{Test 2} + \text{Test 3} + \ldots\text{Test 6} + \text{Test 7}}{\text{\# of Tests Performed during the Seven Days}} \quad (6.1)$$

DOI: 10.1201/9781003354314-6

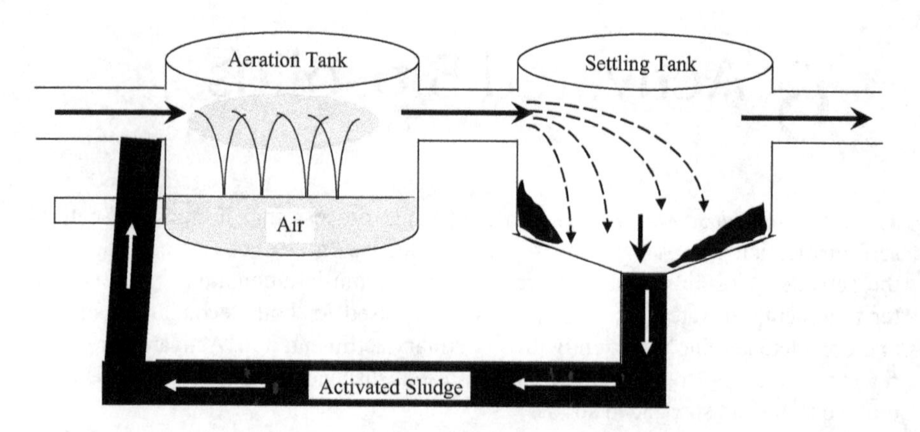

FIGURE 6.1 The activated sludge process.

Example 6.1

Problem:

Calculate the seven-day moving average for days 7, 8, and 9.

Day	MLSS	Day	MLSS
1	3,340	6	2,780
2	2,480	7	2,476
3	2,398	8	2,756
4	2,480	9	2,655
5	2,558	10	2,396

Solution:

(1) $$\text{Moving Ave., Day 7} = \frac{3{,}340 + 2{,}480 + 2{,}398 + 2{,}480 + 2{,}558 + 2{,}780 + 2{,}476}{7} = 2{,}645$$

(2) $$\text{Moving Ave., Day 8} = \frac{2{,}480 + 2{,}398 + 2{,}480 + 2{,}558 + 2{,}780 + 2{,}476 + 2{,}756}{7} = 2{,}561$$

(3) $$\text{Moving Ave., Day 9} = \frac{2{,}398 + 2{,}480 + 2{,}558 + 2{,}780 + 2{,}476 + 2{,}756 + 2{,}655}{7} = 2{,}586$$

6.1.2 BOD or COD Loading

When calculating BOD, COD, or SS loading on an aeration process (or any other treatment process), loading on the process is usually calculated as pounds per day. The following equation is used:

$$\text{BOD, COD, or SS Loading, lb/day} = (\text{mg/L})\ (\text{MGD})\ (8.34\ \text{lb/gal}) \quad (6.2)$$

Example 6.2

Problem:
The BOD concentration of the wastewater entering an aerator is 210 mg/L (see Figure 6.2). If the flow to the aerator is 1,550,000 gpd, what is the pounds per day BOD loading?

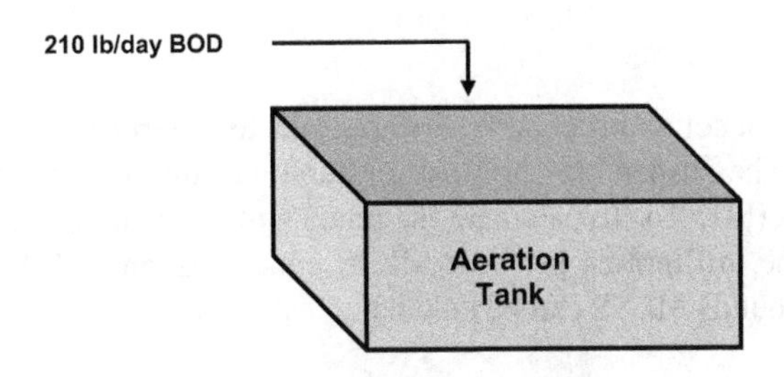

FIGURE 6.2 Refer to example 6.2.

Solution:

$$\begin{aligned}\text{BOD, lbs/day} &= (\text{BOD, mg/L})\ (\text{Flow, MGD})\ (8.34\ \text{lbs/gal})\\ &= (210\ \text{mg/L})\ (1.55\ \text{MGD})\ (8.34\ \text{lbs/gal})\\ &= 2715\ \text{lbs/day}\end{aligned}$$

Example 6.3

Problem:
The flow to an aeration tank is 2,750 gpm. If the BOD concentration of the wastewater is 140 mg/L (see Figure 6.3), how many pounds of BOD are applied to the aeration tank daily?

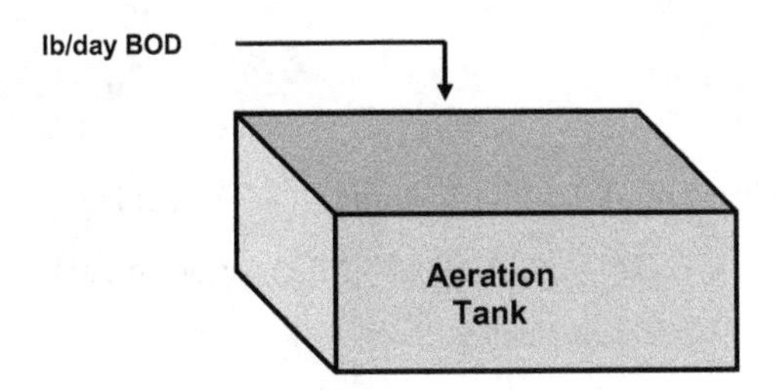

FIGURE 6.3 Refer to example 6.3.

Solution:
First, convert the gallons per minute flow to gallons per day flow:

$$(2750\ \text{gpm})\ (1440\ \text{min/day}) = 3,960,000\ \text{gpd}$$

Then, calculate pounds per day BOD:

$$\begin{aligned}\text{BOD, lb/day} &= (\text{BOD, mg/L})\ (\text{Flow, MGD})\ (8.34\ \text{lb/gal})\\ &= (140,\ \text{mg/L})\ (3.96\ \text{MGD})\ (8.34\ \text{lb/day})\\ &= 4624\ \text{lb/day}\end{aligned}$$

6.1.3 Solids Inventory

In the activated biosolids process, it is important to control the amount of solids under aeration. The suspended solids in an aeration tank are called mixed liquor suspended solids (MLSS). To calculate the pounds of solids in the aeration tank, we need to know the milligrams per liter MLSS concentration and the aeration tank volume. Then, pounds MLSS can be calculated as follows:

$$\text{lb MLSS} = (\text{MLSS, mg/L})\ (\text{MG})\ (8.34) \tag{6.3}$$

Example 6.4

Problem:
If the mixed liquor suspended solids concentration is 1,200 mg/L (see Figure 6.4) and the aeration tanks has a volume of 550,000 gal, how many pounds of suspended solids are in the aeration tank?

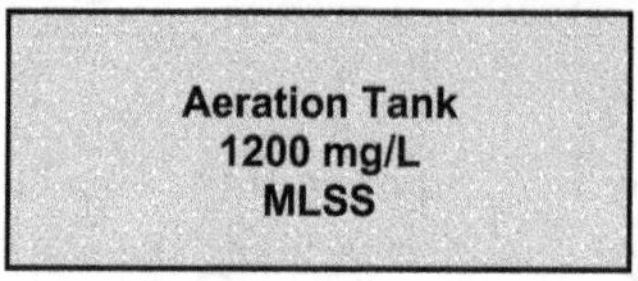

Aeration Tank
1200 mg/L
MLSS

FIGURE 6.4 Refer to example 6.4.

Solution:

$$\begin{aligned}\text{lb} &= (\text{mg/L})\ (\text{MG Volume})\ (8.34\ \text{lb/gal})\\ &= (1200\ \text{mg/L})\ (0.550\ \text{MG})\ (8.34\ \text{lb/gal})\\ &= 5504\ \text{lb MLSS}\end{aligned}$$

6.1.4 Food-to-Microorganism Ratio (F/M Ratio)

The food-to-microorganism ratio (F/M ratio) is a process control method/calculation based upon maintaining a specified balance between available food materials (BOD or COD) in the aeration tank influent and the aeration tank mixed liquor volatile suspended solids (MLVSS) concentration (see Figure 6.5). The chemical oxygen demand (CDO) test is sometimes used, because the results are available in a relatively short period of time. To calculate the F/M ratio, the following information is required:

- Aeration tank influent flow rate, MGD
- Aeration tank influent BOD or COD, mg/L
- Aeration tank MLVSS, mg/L
- Aeration tank volume, MG

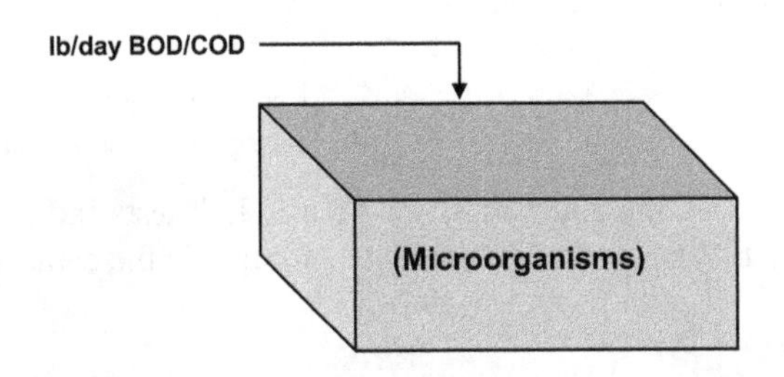

FIGURE 6.5 F/M ratio process control unit.

$$\text{F/M Ratio} = \frac{\text{Primary Eff. COD/BOD mg/L} \times \text{Flow MGD} \times 8.34 \text{ lb/mg/L/MG}}{\text{MLVSS mg/L} \times \text{Aerator Vol., MG} \times 8.34 \text{ lb/mg/L/MG}} \tag{6.4}$$

Typical F/M ratio for activated biosolids process is shown in the following:

Process	Lb BOD / lb MLVSS	lb COD / lb MLVSS
Conventional	0.2–0.4	0.5–1.0
Contact stabilization	0.2–0.6	0.5–1.0
Extended aeration	0.05–0.15	0.2–0.5
Pure oxygen	0.25–1.0	0.5–2.0

Example 6.5

Problem:
The aeration tank influent BOD is 145 mg/L, and the aeration tank influent flow rate is 1.6 MGD. What is the F/M ratio if the MLVSS is 2,300 mg/L and the aeration tank volume is 1.8 MG?

Solution:

$$\text{F/M ratio} = \frac{145 \text{ mg/L} \times 1.6 \text{ MGD} \times 8.34 \text{ lb/mg/L/MG}}{2{,}300 \text{ mg/L} \times 1.8 \text{ MG} \times 8.34 \text{ lb/mg/L/M}}$$

$$= 0.0.6\text{lb BOD/lb MLVSS}$$

✓ **Key Point:** If the MLVSS concentration is not available, it can be calculated if the percent volatile matter (% VM) of the mixed liquor suspended solids (MLSS) is known.

$$\text{MLVSS} = \text{MLSS} \times \% \left(\text{decimal}\right) \text{ Volatile Matter } \left(\text{VM}\right) \tag{6.5}$$

✓ **Key Point:** The "F" value in the F/M ratio for computing loading to an activated biosolids process can be either BOD or COD. Remember, the reason for biosolids production in the activated biosolids process is to convert BOD to bacteria. One advantage of using COD over BOD for analysis of organic load is that COD is more accurate.

Example 6.6

Problem:
The aeration tank contains 2,885 mg/L of MLSS. Lab tests indicate the MLSS is 66% volatile matter. What is the MLVSS concentration in the aeration tank?

Solution:

$$\text{MLVSS, mg/L} = 2{,}885\ \text{mg/L} \times 0.66 = 1{,}904\ \text{mg/L}$$

REQUIRED MLVSS QUANTITY (POUNDS)

The pounds of MLVSS required in the aeration tank to achieve the optimum F/M ration can be determined from the average influent food (BOD or COD) and the desired F/M ratio.

$$\text{MLVSS, lb} = \frac{\text{Primary Effluent BOD or COD} \times \text{Flow, MGD} \times 8.34}{\text{Desired F/M Ratio}} \qquad (6.6)$$

The required pounds of MLVSS determined by this calculation can then be converted to a concentration value by:

$$\text{MLVSS, mg/L} = \frac{\text{Desired MLVSS in Pounds}}{[\text{Aeration Volume, MG} \times 8.34]} \qquad (6.7)$$

Example 6.7

Problem:
The aeration tank influent flow is 4 MGD, and the influent COD is 145 mg/L. The aeration tank volume is 0.65 MG. The desired F/M ratio is 0.3 lb COD/lb MLVSS. How many pounds of MLVSS must be maintained in the aeration tank to achieve the desired F/M ratio? What is the required concentration of MLVSS in the aeration tank?

$$\text{MLVSS} = \frac{145\ \text{mg/L} \times 4.0\ \text{MGD} \times 8.34\ \text{lb/gal}}{[0.3\text{lb COD/lb MLVSS}]} = 16{,}124\ \text{lb MLVSS}$$

$$\text{MLVSS, mg/L} = \frac{16{,}124\ \text{lb MLVSS}}{[0.65\ \text{MG x } 8.34]} = 2{,}974\ \text{mg/L MLVSS}$$

CALCULATING WASTE RATES USING F/M RATIO

Maintaining the desired F/M ratio is accomplished by controlling the MLVSS level in the aeration tank. This may be accomplished by adjustment of return rates; however, the most practical method is by proper control of the waste rate.

$$\text{Waste Vol. Solids, lb/day} = \text{Actual MLVSS, lb} - \text{Desired MLVSS, lb} \tag{6.8}$$

If the desired MLVSS is greater than the actual MLVSS, wasting is stopped until the desired level is achieved. Practical considerations require that the required waste quantity be converted to a required volume to waste per day. This is accomplished by converting the waste pounds to flow rate in million gallons per day or gallons per minute.

$$\text{Waste, MGD} = \frac{\text{Waste Volatile, lb/day}}{[\text{Waste Volatile Conc., mg/L} \times 8.34]} \tag{6.9}$$

$$\text{Waste, gpm} = \frac{\text{Waste, MGD} \times 1{,}000{,}000 \text{ gpd/MGD}}{[1{,}440 \text{ minute/day}]} \tag{6.10}$$

✓ **Key Point:** When F/M ratio is used for process control, the volatile content of the waste activated sludge should be determined.

Example 6.8

Problem:
Given the following information, determine the required waste rate in gallons per minute to maintain an F/M ratio of 0.17 lb COD/lb MLVSS.

Primary effluent COD	140 mg/L
Primary effluent flow	2.2 MGD
MLVSS, mg/L	3,549 mg/L
Aeration tank volume	0.75 MG
Waste volatile concentrations	4,440 mg/L (volatile solids)

$$\text{Actual MLVSS, lb} = 3{,}549 \text{ mg/L} \times 0.75 \text{ MG x } 8.34 = 22{,}199 \text{ lb}$$

$$\text{Required MLVSS, lb} = \frac{140 \text{ mg/L x } 2.2 \text{ MGD x } 8.34}{0.17 \text{lb COD/lb MLVSS}} 15{,}110 \text{ lb MLVSS}$$

$$\text{Waste, lb/day} = 22{,}199 \text{ lb} - 15{,}110 \text{ lb} = 7{,}089 \text{ lb}$$

$$\text{Waste, MGD} = \frac{7{,}089 \text{ lb/day}}{4{,}440 \text{ mg/L x } 8.34} = 0.19 \text{ MGD}$$

$$\text{Waste, gpm} = \frac{0.19 \text{ MGD} \times 1{,}000{,}000 \text{ gpd/MGD}}{1{,}440 \text{ min/day}} = 132 \text{ gpm}$$

6.1.5 Gould Biosolids Age

Biosolids age refers to the average number of days a particle of suspended solids remains under aeration. It is a calculation used to maintain the proper amount of activated biosolids in the aeration tank. This calculation is sometimes referred to as Gould biosolids age so that it is not confused with similar calculations, such as solids retention time (or mean cell residence time).

When considering sludge age, in effect, we are asking how many days of suspended solids are in the aeration tank. For example, if 3,000 lb SS enter the aeration tank daily and the aeration tank contains 12,000 lb of suspended solids, when 4 days of solids are in the aeration tank, the sludge age is 4 days.

$$\text{Sludge Age, days} = \frac{\text{SS in Tank, lb}}{\text{SS Added, lb/day}} \tag{6.11}$$

Example 6.9

Problem:
A total of 2,740 lb/day suspended solids enters an aeration tank in the primary effluent flow. If the aeration tank has a total of 13,800 lb of mixed liquor suspended solids, what is the biosolids age in the aeration tank?

Solution:
Referring to equation 6.11:

$$= \frac{13{,}800 \text{ lb}}{2740 \text{ lb/day}}$$

$$= 5.0 \text{ days}$$

6.1.6 Mean Cell Residence Time (MCRT)

Mean cell residence time (MCRT), sometimes called *sludge retention time*, is another process control calculation used for activated biosolids systems. MCRT represents the average length of time an activated biosolids particle remains in the activated biosolids system. It can also be defined as the length of time required at the current removal rate to remove all the solids in the system.

Mean cell residence time, day:

$$\frac{\left[\text{MLSS mg/L} \times (\text{Aeration Vol.} + \text{Clarifier Vol.}) \times 8.34 \text{ lb/mg/L/MG}\right]}{\left[\left[\text{WAS, mg/L} \times \text{WAS flow x } 8.34\right) + (\text{TSS out} \times \text{flow out} \times 8.34)\right]} \tag{6.12}$$

✓ **Key Point:** MCRT can be calculated using only the aeration tank solids inventory. When comparing plant operational levels to reference materials, you must determine which calculation the reference manual uses to obtain its example values. Other methods are available to determine the clarifier solids concentrations. However, the simplest method assumes that the average suspended solids concentration is equal to the aeration tank's solids concentration.

Example 6.10

Problem:
Given the following data, what is the MCRT?

Aerator volume	= 1,000,000 gal
Final clarifier	= 600,000 gal
Flow	= 5.0 MGD
Waste rate	= 0.085 MGD
MLSS, mg/L	= 2,500 mg/L
Waste, mg/L	= 6,400 mg/L
Effluent, TSS	= 14 mg/L

Solution:

$$\text{MRCT} = \frac{[2{,}500\text{ mg/L} \times (1.0\text{ MG} + 0.60\text{ MG}) \times 8.34]}{[6{,}4000\text{ mg/L} \times 0.085\text{ MGDx}8.34) + (14\text{ mg/L} \times 5.0\text{ mgd} \times 8.34)]} = 6.5\text{ days}$$

WASTE QUANTITIES/REQUIREMENTS

MCRT for process control requires the determination of the optimum range for MCRT values. This is accomplished by comparison of the effluent quality with MCRT values. When the optimum MCRT is established, the quantity of solids to be removed (wasted) is determined by:

$$\text{Waste, lb/day} = \left(\frac{\text{MLSS} \times (\text{Aer., MG} + \text{Clarifier, MG}) \times 8.34}{\text{Desired MCRT}}\right) - [\text{TSS}_{\text{out}} \times \text{Flow} \times 8.34] \quad (6.13)$$

Example 6.11

$$\frac{3{,}400\text{ mg/L} \times (1.4\text{ MG} + 0.50\text{ MG}) \times 8.34}{8.6\text{ days}} - [10\text{ mg/L} \times 5.0\text{ MGD} \times 8.34]$$

$$\text{Waste Quality, lb/day} = 5{,}848\text{ lb}$$

6.1.7 Waste Rate in Million Gallons/Day

When the quantity of solids to be removed from the system is known, the desired waste rate in million gallons per day can be determined. The unit used to express the rate (MGD, gallons per day, and gallons per minute) is a function of the volume of waste to be removed and the design of the equipment.

$$\text{Waste, MGD} = \frac{\text{Waste Pounds/day}}{\text{WAS Concentration, mg/L} \times 8.34} \quad (6.14)$$

$$\text{Waste, gpm} = \frac{\text{Waste MGD} \times 1{,}000{,}000\text{ gpd/MGD}}{1.440\text{ minutes/day}} \quad (6.15)$$

Example 6.12

Problem:
Given the following data, determine the required waste rate to maintain an MCRT in 8.8 days.

MLSS, mg/L	2,500 mg/L
Aeration volume	1.20 MG
Clarifier volume	0.20 MG
Effluent TSS	11 mg/L
Effluent flow	5.0 MGD
Waste concentration	6,000 mg/L

Solution:

$$\text{Waste, lb/day} = \frac{2{,}500\ \text{mg/L} \times (1.20 + 0.20) \times 8.34}{8.8\ \text{days}} - [11\ \text{mg/L} \times 5.0\ \text{MGD} \times 8.34]$$

$$= 3{,}317\ \text{lb/day} - 459\ \text{lb/day}$$

$$2{,}858\ \text{lb/day}$$

$$\text{Waste, lb/day} = \frac{2{,}858\ \text{lb/day}}{[6{,}000\ \text{mg/L} \times 8.34]} = 0.057\ \text{MGD}$$

$$\text{Waste, gpm} = \frac{0.057\ \text{MGD} \times 1{,}000{,}000\text{gpd/MGD}}{1{,}440\ \text{min/day}} = 40\ \text{gpm}$$

6.1.8 Estimating Return Rates from SSV_{60}

Many methods are available for estimation of the proper return biosolids rate. A simple method described in the *Operation of Wastewater Treatment Plants, Field Study Programs* (1986), developed by the California State University, Sacramento, uses the 60 min percent settled sludge volume. The $\%SSV_{60}$ test results can provide an approximation of the appropriate return activated biosolids rate. This calculation assumes that the SSV_{60} results are representative of the actual settling occurring in the clarifier. If this is true, the return rate in percent should be approximately equal to the SSV_{60}. To determine the approximate return rate in million gallons per day (MGD), the influent flow rate, the current return rate, and the SSV_{60} must be known. The results of this calculation can then be adjusted based upon sampling and visual observations to develop the optimum return biosolids rate.

✓ **Key Point:** The percent SSV_{60} must be converted to a decimal percent and total flow rate (wastewater flow and current return rate in million gallons per day must be used).

$$\text{Est. Return Rate, MGD} = (\text{Influent Flow, MGD} + \text{Current Return Flow, MGD}) \times \%SSV_{60} \quad (6.16)$$

$$\text{RAS Rate, GPM} = \frac{\text{Return, Biosolids Rate, gpd}}{1{,}440\ \text{min/day}} \quad (6.17)$$

Assume:

- Percent SSV_{60} is representative.
- Return rate in percent equals percent SSV_{60}.
- Actual return rate is normally set slightly higher to ensure organisms are returned to the aeration tank as quickly as possible. The rate of return must be adequately controlled to prevent the following:
    - Aeration and settling hydraulic overloads
    - Low MLSS levels in the aerator
    - Organic overloading of aeration
    - Septic return activated biosolids
    - Solids loss due to excessive biosolids blanket depth

Example 6.13

Problem:
The influent flow rate is 5 MGD, and the current return activated sludge flow rate is 1.8 MGD. The SSV_{60} is 37%. Based upon this information, what should be the return biosolids rate in million gallons per day (MGD)?

Solution:

$$\text{Return, MGD} = (5.0\text{ MGD} + 1.8\text{ MGD}) \times 0.37 = 2.5\text{ MGD}$$

6.1.9 Sludge Volume Index (SVI)

Sludge volume index (SVI) is a measure (an indicator) of the settling quality (a quality indicator) of the activated biosolids. As the SVI increases, the biosolids settles slower, do not compact as well, and are likely to result in an increase in effluent suspended solids. As the SVI decreases, the biosolids become denser, settling is more rapid, and the biosolids age. SVI is the volume in milliliters occupied by 1 g of activated biosolids. For the settled biosolids volume (mL/L) and the mixed liquor suspended solids (MLSS) calculation, milligrams per liter are required. The proper SVI range for any plant must be determined by comparing SVI values with plant effluent quality.

$$\text{Sludge Volume Index (SVI)} = \frac{\text{SSV, mL/L} \times 1{,}000}{\text{MLSS, mg/L}} \tag{6.18}$$

Example 6.14

Problem:
The SSV_{30} is 365 mL/L, and the MLSS is 2,365 mg/L. What is the SVI?

Solution:

$$\text{Sludge Volume Index (SVI)} = \frac{365\text{ mL/L} \times 1{,}000}{2{,}365\text{ mg/L}} = 154.3$$

SVI equals 154.3. What does this mean? It means that the system is operating normally, with good settling and low effluent turbidity. How do we know this? We

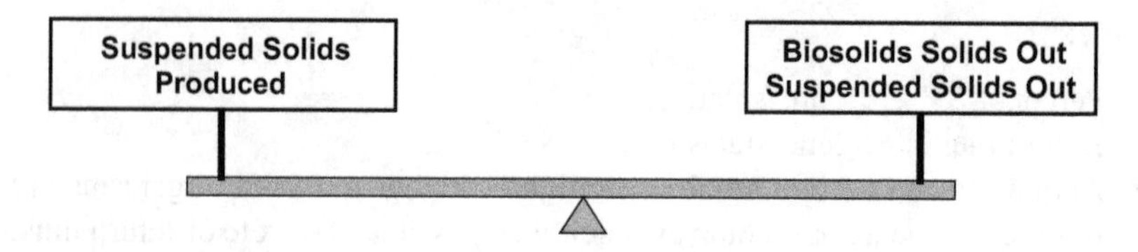

FIGURE 6.6 Biological process mass balance.

know this because we compare the 154.3 result with the parameters listed in what follows to obtain the expected condition (the result).

SVI	Expected Conditions (Indicates)
Less than 100	Old biosolids—possible pin floc Effluent turbidity increasing
100–250	Normal operation—good settling Low effluent turbidity
Greater than 250	Bulking biosolids—poor settling High effluent turbidity

6.1.10 Mass Balance: Settling Tank Suspended Solids

Solids are produced whenever biological processes are used to remove organic matter from wastewater (see Figure 6.6). Mass balance for anaerobic biological process must consider both the solids removed by physical settling processes and the solids produced by biological conversion of soluble organic matter to insoluble suspended matter organisms. Research has shown that the amount of solids produced per pound of BOD removed can be predicted based upon the type of process being used. Although the exact amount of solids produced can vary from plant to plant, research has developed a series of *K* factors that can be used to estimate the solids production for plants using a particular treatment process. These average factors provide a simple method to evaluate the effectiveness of a facility's process control program. The mass balance also provides an excellent mechanism to evaluate the validity of process control and effluent monitoring data generated.

<u>MASS BALANCE CALCULATION</u>

$$\text{BOD in, lb} = \text{BOD, mg/L} \times \text{flow, MGD} \times 8.34 \tag{6.19}$$

$$\text{BOD out, lb} = \text{BOD, mg/L} \times \text{flow, MGD} \times 8.34$$

$$\text{solids produced, lb/day} = (\text{BOD in, lb} - \text{BOD out, lb}) \times \text{K}$$

$$\text{TSS out, lb/day} = \text{TSS out, mg/L} \times \text{flow, MGD} \times 8.34$$

$$\text{waste, lb/day} = \text{waste, mg/L} \times \text{flow, MGD} \times 8.34$$

solids removed, lb/day = TSS out, lb/day + waste, lb/day

$$\% \text{ Mass Balance} = \frac{(\text{Solids Produced} - \text{Solids Removed}) \times 100}{\text{Solids Produced}}$$

6.1.11 Biosolids Waste Based upon Mass Balance

$$\text{Waste Rate, MGD} = \frac{\text{Solids Produced, lb/day}}{(\text{Waste Concentration} \times 8.34)} \tag{6.20}$$

Example 6.15

Problem:
Given the following data, determine the mass balance of the biological process and the appropriate waste rate to maintain current operating conditions.

Process	Extended Aeration (No Primary)	
Influent	Flow	1.1 MGD
	BOD	220 mg/L
	TSS	240 mg/L
Effluent	Flow	1.5 MGD
	BOD	18 mg/L
	TSS	22 mg/L
Waste	Flow	24,000 gpd
	TSS	8,710 mg/L

Solution:

BOD in = 220 mg/L × 1.1 MGD × 8.34= 2,018 lb/day

BOD out = 18 mg/L × 1.1 MGD × 8.34 = 165 lb/day

BOD removed = 2,018 lb/day − 165 lb/day = 1,853 lb/day

solids produced = 1,853 lb/day × 0.65 lb/lb BOD = 1,204 lb solids/day

solids out, lb/day = 22 mg/L × 1.1 MGD × 8.34 = 202 lb/day

sludge out, lb/day = 8,710 mg/L × 0.024 MGD × 8.34 = 1,743 lb/day

solids removed, lb/day = (202 lb/day + 1,743 lb/day) =1,945 lb/day

$$\text{Mass Balance} = \frac{(1{,}204 \text{ lb Solids/day} - 1{,}945\text{lb/day}) \times 100}{1{,}204 \text{ lb/day}} = \text{-}62\%$$

The mass balance indicates:

(1) The sampling point(s), collection methods, and/or laboratory testing procedures are producing non-representative results.
(2) The process is removing significantly more solids than is required. Additional testing should be performed to isolate the specific cause of the imbalance.

To assist in the evaluation, the waste rate based upon the mass balance information can be calculated.

$$\text{Waste, GPD} = \frac{\text{Solids Produced, lb/day}}{(\text{Waste TSS, mg/L x 8.34})} \tag{6.21}$$

$$\text{Waste, GPD} = \frac{1{,}204 \text{ lb/day} \times 1{,}000{,}000}{8{,}710 \text{ mg/L} \times 8.34} = 16{,}575 \text{ gpd}$$

6.1.12 Oxidation Ditch Detention Time

Oxidation ditch systems may be used where the treatment of wastewater is amendable to aerobic biological treatment and the plant design capacities generally do not exceed 1.0 MGD. The oxidation ditch is a form of aeration basin where the wastewater is mixed with return biosolids (see Figure 6.7). The oxidation ditch is essentially a modification of a completely mixed activated biosolids system used to treat wastewater from small communities. This system can be classified as an extended aeration process and is considered to be a low–loading rate system. This type of treatment facility can remove 90% or more of influent BOD. Oxygen requirements will generally depend on the maximum diurnal organic loading, degree of treatment, and suspended solids concentration to be maintained in the aerated channel mixed liquor suspended solids (MLSS). *Detention time* is the length of time required for a given flow rate to pass through a tank. Detention time is not normally calculated for aeration basins, but it is calculated for oxidation ditches.

✓ **Key Point:** When calculating detention time, it is essential that the time and volume units used in the equation are consistent with each other.

$$\text{Detention Time, hrs} = \frac{\text{Vol. of Oxidation Ditch, gal}}{\text{Flow Rate, gph}} \tag{6.22}$$

Example 6.16

Problem:
An oxidation ditch has a volume of 160,000 gal. If the flow to the oxidation ditch is 185,000 gpd, what is the detention time in hours?

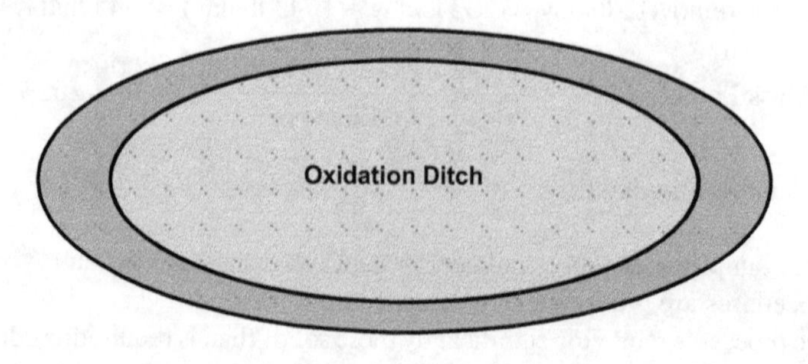

FIGURE 6.7 Oxidation ditch.

Solution:

Because detention time is desired in hours, the flow must be expressed as gallons per hour:

$$\frac{185{,}000 \text{ gpd}}{24 \text{ hrs/day}} = 7708 \text{ gph}$$

Now, calculate detention time:

$$\text{Detention Time, hrs} = \frac{\text{Vol. of Oxidation Ditch, gal}}{\text{Flow Rate, gph}}$$

$$= \frac{160{,}000 \text{ gallons}}{7708 \text{ gph}}$$

$$= 20.8 \text{ hrs}$$

6.2 ACTIVATED BIOSOLIDS PRACTICE PROBLEMS

Problem 6.1

The flow to an aeration tank is 900,000 gpd. If the BOD content of the wastewater entering the aeration tanks is 220 mg/L, how many pounds of BOD are applied to the aeration tanks daily?

Solution:

$$\text{lb BOD/day} = (\text{BOD, mg/L})(\text{Flow, MGD})(8.34)$$

$$= (220 \text{ mg/L})(0.9 \text{ MG})(8.34)$$

$$= 1651 \text{ lb/d}$$

Problem 6.2

The flow to an aeration tank is 1,100 gpm. If the COD concentration of the wastewater is 150 mg/L, what is the COD loading rate in pounds per day?

Solution:

$$\frac{(1100 \text{ gpm})(1440)}{1{,}000{,}000} = 1.584 \text{ MGD}$$

$$\text{lb COD/day} = (\text{COD, mg/L})(\text{Flow, MGD})(8.34)$$

$$= (150 \text{ mg/L})(1.584 \text{ MGD})(8.34)$$

$$= 1982 \text{ lb/d}$$

Problem 6.3

An aeration basin is 110 ft long, 40 ft wide and holds wastewater to a depth of 10 ft. If the aeration basin has an MLSS concentration of 2,000 mg/L, how many pounds of MLSS are under aeration?

Solution:

$$\text{Aerator volume, gal} = (110\text{ ft})(40\text{ ft})(10\text{ ft})(7.48) = 329{,}120\text{ gal}$$

$$\begin{aligned}\text{MLSS, lb} &= (\text{MLSS, mg/L})(\text{Aerator Vol, MG})(8.34)\\ &= (2000\text{ mg/L})(0.329{,}120\text{ MG})(8.34)\\ &= 5490\text{ lb MLSS}\end{aligned}$$

Problem 6.4

The aeration tank of a conventional activate biosolids plant has an MLSS concentration of 2,200 mg/L with a volatile solids content of 70%. If the volume of the aeration tank is 210,000 gal, how many pounds of volatile solids are under aeration?

Solution:

$$\begin{aligned}\text{MLVSS, lb} &= (\text{MLSS, mg/L})(\text{Aerator Volume, MG})(8.34)(\%\text{ VS})\\ &= (2200\text{ mg/L})(0.21\text{ MG})(8.34)(0.70)\\ &= 2697\text{ lb MLVSS}\end{aligned}$$

Problem 6.5

An activated biosolids aeration tank receives a primary effluent flow of 1.5 MGD, with a BOD concentration of 170 mg/L. The mixed liquor volatile suspended solids is 2,100 mg/L, and the aeration tank volume is 400,000 gallon. What is the current F/M ratio?

Solution:

$$\begin{aligned}\text{F/M} &= \frac{(\text{BOD or COD, mg/L})(\text{Flow, MGD})(8.34)}{(\text{MLVSS, mg/L})(\text{Aerator Volume, MG})(8.34)}\\ &= \frac{(170\text{ mg/L})(1.5\text{ MGD})(8.34)}{(2100\text{ mg/L})(0.4\text{ MG})(8.34)} = \frac{2126.7}{7005.6} = 0.30\end{aligned}$$

7 Treatment Ponds

The primary goals of wastewater treatment ponds focus on simplicity and flexibility of operation, protection of the water environment, and protection of public health. Moreover, ponds are relatively easy to build and manage, they accommodate large fluctuations in flow, and they can also provide treatment that approaches conventional systems (producing a highly purified effluent) at a much lower cost. It is the cost (the economics) that drives many managers to decide on the pond option of treatment. The actual degree of treatment provided in a pond depends on the type and number of ponds used. Ponds can be used as the sole type of treatment, or they can be used in conjunction with other forms of wastewater treatment—that is, other treatment processes followed by a pond, or a pond followed by other treatment processes. Ponds can be classified based upon their location in the system, by the type of wastes they receive, and by the main biological process occurring in the pond.

7.1 TREATMENT POND PARAMETERS

Before we discuss process control calculations, it is important first to describe the calculations used for determining the area, volume, and flow rate parameters that are crucial in making treatment pond calculations.

7.1.1 Determining Pond Area (Inches)

$$\text{Area, acres} = \frac{\text{Area, ft}^2}{43{,}560\ \text{ft}^2\text{/acre}} \tag{7.1}$$

7.1.2 Determining Pond Volume (Acre-Feet)

$$\text{Volume, acre-feet} = \frac{\text{Volume, ft}^3}{43{,}560\ \text{ft}^2\text{/acre-foot}} \tag{7.2}$$

7.1.3 Determining Flow Rate in Acre-Feet/Day

$$\text{Flow, acre-feet/day} = \text{flow, MGD} \times 3{,}069\ \text{acre-feet/MG} \tag{7.3}$$

✓ **Key Point:** Acre-feet (ac-ft) is a unit that can cause confusion, especially for those not familiar with pond or lagoon operations; 1 ac-ft is the volume of a box with a 1 ac top and 1 ft of depth, but the top doesn't have to be an even number of acres in size to use acre-feet.

DOI: 10.1201/9781003354314-7

7.1.4 Determining Flow Rate in Acre-Inches/Day

$$\text{Flow, acre-inches/day} = \text{flow, MGD} \times 36.8 \text{ acre-inches/MG} \tag{7.4}$$

7.2 TREATMENT POND PROCESS CONTROL CALCULATIONS

Although there are no recommended process control calculations for the treatment pond, there are several calculations that may be helpful in evaluating process performance or identifying causes of poor performance. These include hydraulic detention time, BOD loading, organic loading rate, BOD removal efficiency, population loading, and hydraulic loading rate. In the following, we provide a few calculations that might be helpful in pond performance evaluation and identification of causes of poor performance process along with other calculations and/or equations that may be helpful.

7.2.1 Hydraulic Detention Time, Days

$$\text{Hydraulic detention time, days} = \frac{\text{Pond volume, acre-feet}}{\text{Influent flow, acre-feet/day}} \tag{7.5}$$

✓ **Key Point:** Normally, hydraulic detention time ranges from 30 to 120 days for stabilization ponds.

Example 7.1

Problem:
A stabilization pond has a volume of 54.5 ac-ft. What is the detention time in days when the flow is 0.35 MGD?

Solution:

$$\text{Flow, ac-ft/day} = 0.35 \text{ MGD} \times 3.069 \text{ ac-ft/MG}$$

$$= 1.07 \text{ ac-ft/day}$$

$$\text{DT days} = \frac{54.5 \text{ acre/ft}}{1.07 \text{ ac-ft/day}}$$

7.2.2 BOD Loading

When calculating BOD loading on a wastewater treatment pond, the following equation is used:

$$\text{lb/day} = (\text{BOD, mg/L})\ (\text{flow, MGD})\ (8.34\text{lb/gal}) \tag{7.6}$$

Example 7.2

Problem:
Calculate the BOD loading (pounds per day) on a pond if the influent flow is 0.3MGD with a BOD of 200 mg/L.

Solution:

$$\begin{aligned}\text{lb/day} &= (\text{BOD, mg/L})\ (\text{flow, MGD})\ (8.34\ \text{lb/gal}) \\ &= (200\ \text{mg/L})\ (0.3\ \text{MGD})\ (8.34\ \text{lb/gal}) \\ &= 500\ \text{lb/day BOD}\end{aligned}$$

7.2.3 Organic Loading Rate

Organic loading can be expressed as pound of BOD per acre per day (most common), pounds of BOD per acre-foot per day, or people per acre per day.

$$\text{Organic Loading, lb BOD/acre/day} = \frac{\text{BOD, mg/L Influ. Flow, MGD} \times 8.34}{\text{Pond area, acres}} \tag{7.7}$$

✓ **Key Point:** Normal range is 10–50 lb BOD per day per acre.

Example 7.3

Problem:
A wastewater treatment pond has an average width of 370 ft and an average length of 730 ft. The influent flow rate to the pond is 0.10 MGD, with a BOD concentration of 165 mg/L. What is the organic loading rate to the pond in pounds per day per acre (lb/d/ac)?

Solution:

$$730\ \text{ft} \times 370\ \text{ft} \times \frac{1\ \text{ac}}{43{,}560\ \text{ft}^2} = 6.2\ \text{acre}$$

$$0.10\ \text{MGD} \times 165\ \text{mg/L} \times 8.34\ \text{lb/gal} = 138\ \text{lb/day}$$

$$\frac{138\ \text{lb/d}}{6.2\ \text{ac}} = 22.2\ \text{lb/d/ac}$$

7.2.4 BOD Removal Efficiency

As mentioned, the efficiency of any treatment process is its effectiveness in removing various constituents from the water or wastewater. BOD removal efficiency is therefore a measure of the effectiveness of the wastewater treatment pond in removing BOD from the wastewater.

$$\%\ \text{BOD Removed} = \frac{\text{BOD Removed, mg/L}}{\text{BOD Total, mg/L}} \times 100$$

Example 7.4

Problem:
The BOD entering a waste treatment pond is 194 mg/L. If the BOD in the pond effluent is 45 mg/L, what is the BOD removal efficiency of the pond?

$$\% \text{ BOD Removed} = \frac{\text{BOD Removed, mg/L}}{\text{BOD Total, mg/L}} \times 100$$

$$= \frac{149\text{mg/L}}{194 \text{ mg/L}} \times 100$$

$$= 77\%$$

7.2.5 Population Loading

$$\text{Population loading, people/acre/day} = \frac{\text{BOD, mg/L Infl.flow, MGD} \times 8.34}{\text{Pond area, acres}} \quad (7.8)$$

$$\text{Hydraulic Loading, inches/day} = \frac{\text{Influent flow, acre-inches/day}}{\text{Pond area, acres}} \quad (7.9)$$

7.3 TREATMENT POND PRACTICE PROBLEMS

Problem 7.1

A stabilization pond has a volume of 52.5 ac-ft. What is the detention time in days when the flow is 0.4 MGD?

Solution:

$$\text{Flow, ac-ft/day} = 0.4 \text{ MGD} \times 3.069 \text{ ac-ft/MG}$$

$$= 1.23 \text{ ac-ft/day}$$

$$\text{DT days} = \frac{52.5 \text{ acre/ft}}{1.23 \text{ ac-ft/day}} = 42.7$$

Problem 7.2

Calculate the BOD loading (pounds per day) on a pond if the influent flow is 0.4 MGD, with a BOD of 210 mg/L.

Solution:

$$\text{lb/day} = (\text{BOD, mg/L})\ (\text{flow, MGD})\ (8.34 \text{ lb/gal})$$

$$= (210 \text{ mg/L})\ (0.4 \text{ MGD})\ (8.34 \text{ lb/gal})$$

$$= 700.6 \text{ lbs/day BOD}$$

Problem 7.3

A wastewater treatment pond has an average width of 350 ft and an average length of 720 ft. The influent flow rate to the pond is 0.10 MGD, with a BOD concentration of 160 mg/L. What is the organic loading rate to the pond in pounds per day per acre (lb/d/ac)?

Solution:

$$720\ \text{ft} \times 350\ \text{ft} \times \frac{1\ \text{ac}}{43{,}560\ \text{ft}^2} = 5.8\ \text{acre}$$

$$0.10\ \text{MGD} \times 160\ \text{mg/L} \times 8.34\ \text{lb/gal} = 133.4\ \text{lb/day}$$

$$\frac{133.4\ \text{lb/d}}{5.8\ \text{ac}} = 23\ \text{lb/d/ac}$$

Problem 7.4

The BOD entering a waste treatment pond is 190 mg/L. If the BOD in the pond effluent is 40 mg/L, what is the BOD removal efficiency of the pond?

$$\%\ \text{BOD Removed} = \frac{\text{BOD Removal, mg/L}}{\text{BOD Total, mg/L}} \times 100$$

$$= \frac{150\ \text{mg/L}}{190\ \text{mg/L}} \times 100$$

$$= 79\%$$

8 Chemical Dosage Calculations

(Note: Earlier we discussed calculations used in the chlorination process. Thus, there is a certain amount of crossover of information presented in this chapter with similar information previously presented. That is, in this chapter we also discuss chlorination, but as it relates to wastewater treatment.)

8.1 CHEMICAL DOSING

Chemicals are used extensively in wastewater treatment (and water treatment) operations. Plant operators add chemicals to various unit processes for slime growth control, corrosion control, odor control, grease removal, BOD reduction, pH control, biosolids bulking control, ammonia oxidation, bacterial reduction, and for other reasons. In order to apply any chemical dose correctly, it is important to make certain dosage calculations. One of the most frequently used calculations in wastewater/water mathematics is the dosage or loading. The general types of milligrams per liter to pounds per day or pounds calculations are for chemical dosage, BOD, COD, SS loading/removal, pounds of solids under aeration, and WAS pumping rate. These calculations are usually made using either equation 8.1 or equation 8.2.

$$(\text{Chemical, mg/L})\ (\text{MGD flow})\ (8.34\ \text{lb/gal}) = \text{lb/day} \tag{8.1}$$

$$(\text{Chemical, mg/L})\ (\text{MG volume})\ (8.34\ \text{lb/gal}) = \text{lb} \tag{8.2}$$

- ✓ **Key Point:** If milligrams per liter concentration represents a concentration in a flow, then million gallons per day (MGD) flow is used as the second factor. However, if the concentration pertains to a tank or pipeline volume, then million gallons (MG) volume is used as the second factor.
- ✓ **Key Point:** Typically, especially in the past, the expression parts per million (ppm) was used as an expression of concentration, because 1 mg/L = 1 ppm. However, current practice is to use milligrams per liter as the preferred expression of concentration.

8.1.1 CHEMICAL FEED RATE

In chemical dosing, a measured amount of chemical is added to the wastewater (or water). The amount of chemical required depends on the type of chemical

DOI: 10.1201/9781003354314-8

used, the reason for dosing, and the flow rate being treated. The two expressions most often used to describe the amount of chemical added or required are:

- Milligrams per liter (mg/L)
- Pounds per day (lb/day)

A milligram per liter is a measure of concentration. For example, consider Figure 8.1a and Figure 8.1b. In this figure, it is apparent that the milligrams per liter concentration expresses a ratio of the milligram chemical in each liter of water. As shown in the following, if a concentration of 5 mg/L is desired, then a total of 15 mg chemical would be required to treat 3 L:

$$\frac{5 \text{ mg} \times 3}{\text{L} \times 3} = \frac{15 \text{ mg}}{3 \text{ L}}$$

The amount of chemical required therefore depends on two factors:

- The desired concentration (milligrams per liter)
- The amount of wastewater to be treated (normally expressed as million gallons per day).

To convert from milligrams per liter to pounds per day, use equation 8.1.

Example 8.1

Problem:

Determine the chlorinator setting (pounds per day) needed to treat a flow of 5 MGD with a chemical dose of 3 mg/L.

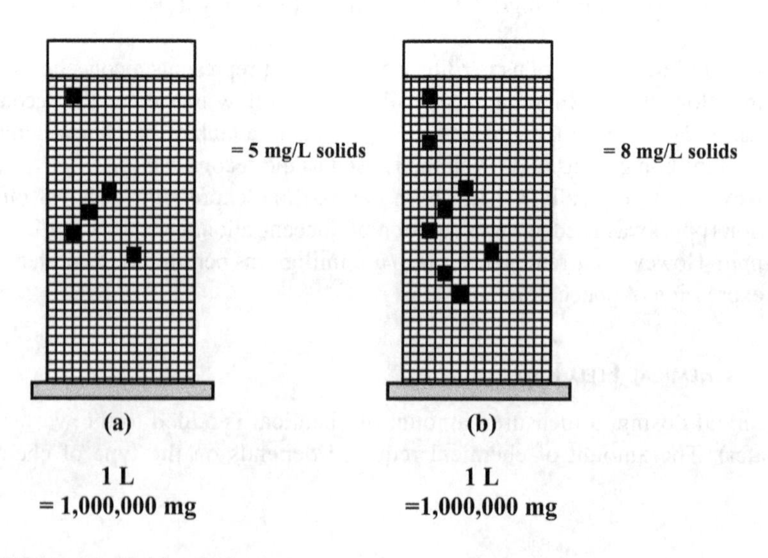

FIGURE 8.1 (a–b) 1 L solutions.

Solution:

$$\text{Chemical, lb/day} = \text{Chemical, mg/L} \times \text{flow, MGD} \times 8.34 \text{ lb/gal}$$
$$= 3 \text{ mg/l} \times 5 \text{ MGD} \times 8.34 \text{ lb/gal}$$
$$= 125 \text{ lb/day}$$

Example 8.2

Problem:
The desired dosage for a dry polymer is 10 mg/L. If the flow to be treated is 2,100,000 gpd, how many pounds per day polymer will be required?

Solution:

$$\text{Polymer, lb/day} = \text{Polymer, mg/L} \times \text{flow, MGD} \times 8.34 \text{ lb/gal}$$
$$= 10 \text{ mg/L Polymer} \times (2.10 \text{ MGD})\ (8.34 \text{ lb/day})$$
$$= 175 \text{ lb/day Polymer}$$

✓ **Key Point:** To calculate chemical dose for tanks or pipelines, a modified equation must be used. Instead of million gallons per day flow, MG volume is used:

$$\text{lb Chemical} = \text{Chemical, mg/L} \times \text{Tank Volume, MG} \times 8.34 \text{ lb/gal} \tag{8.3}$$

Example 8.3

Problem:
To neutralize a sour digester, 1 lb of lime is added for every pound of volatile acids in the digester biosolids. If the digester contains 300,000 gal of biosolids with a volatile acid (VA) level of 2,200 mg/L, how many pounds of lime should be added?

Solution:
Because volatile acid concentration is 2,200 mg/L, the lime concentration should also be 2,200 mg/L:

$$\text{lb lime required} = \text{lime, mg/L} \times \text{digester volume, MG} \times 8.34 \text{ lb/gal}$$
$$= (2200 \text{ mg/L})\ (0.30 \text{ MG})\ 8.34 \text{ lb/gal})$$
$$= 5{,}504 \text{ lb lime}$$

8.1.2 Chlorine Dose, Demand, and Residual

Chlorine is a powerful oxidizer that is commonly used in wastewater and water treatment for disinfection, in wastewater treatment for odor control, bulking control, and other applications. When chlorine is added to a unit process, we want to ensure that a measured amount is added, obviously. Chlorine dose depends on two considerations—the chlorine demand and the desired chlorine residual:

$$\text{Chlorine Dose} = \text{Chlorine Demand} + \text{Chlorine Residual} \tag{8.4}$$

8.1.2.1 Chlorine Dose

In describing the amount of chemical added or required, we use equation (8.1):

$$\text{lb/day} = \text{Chemical, mg/L} \times \text{MGD} \times 8.34\text{, lb/day} \tag{8.5}$$

Example 8.4

Problem:
Determine the chlorinator setting (pounds per day) needed to treat a flow of 8 MGD with a chlorine dose of 6 mg/L.

Solution:

$$(\text{mg/L})(\text{MGD})(8.34) = \text{lb/day}$$
$$(6\text{ mg/L})\ (8\text{ MGD})\ (8.34\text{ lb/gal}) = \text{lb/day}$$
$$= 400\text{ lb/day}$$

8.1.2.2 Chlorine Demand

The *chlorine demand* is the amount of chlorine used in reacting with various components of the water, such as harmful organisms and other organic and inorganic substances. When the chlorine demand has been satisfied, these reactions cease.

Example 8.5

Problem:
The chlorine dosage for a secondary effluent is 6 mg/L. If the chlorine residual after 30 min contact time is found to be 0.5 mg/L, what is the chlorine demand expressed in milligrams per liter?

Solution:

$$\text{Chlorine Dose} = \text{Chlorine demand} + \text{Chlorine Residual}$$
$$6\text{ mg/L} = x\text{mg/L} + 0.5\text{ mg/L}$$
$$6\text{ mg/L} - 0.5\text{ mg/L} = x\text{mg/L}$$
$$x = 5.5\text{ mg/L Chlorine Demand}$$

8.1.2.3 Chlorine Residual

Chlorine residual is the amount of chlorine remaining after the demand has been satisfied.

Example 8.6

Problem:
What should the chlorinator setting be (pounds per day) to treat a flow of 3.9 MGD if the chlorine demand is 8 mg/L and a chlorine residual of 2 mg/L is desired?

Solution:
First, calculate the chlorine dosage in milligrams per liter:

$$\begin{aligned}\text{Chlorine Dose} &= \text{Chlorine demand} + \text{Chlorine Residual}\\ &= 8 \text{ mg/L} + 2 \text{ mg/L}\\ &= 10 \text{ mg/L}\end{aligned}$$

Then, calculate the chlorine dosage (feed rate) in pounds per day:

$$(\text{Chlorine, mg/L})\ (\text{MGD flow})\ (8.34 \text{ lb/gal}) = \text{lb/day Chlorine}$$

$$(10 \text{ mg/L})\ (3.9 \text{ MGD})\ (8.34 \text{ lb/gal}) = 325 \text{ lb/day Chlorine}$$

8.1.3 Hypochlorite Dosage

Hypochlorite is less hazardous than chlorine; therefore, it is often used as a substitute chemical for elemental chlorine. Hypochlorite is similar to strong bleach and comes in two forms: dry calcium hypochlorite (often referred to as HTH) and liquid sodium hypochlorite. Calcium hypochlorite contains about 65% available chlorine; sodium hypochlorite contains about 12% to 15% available chlorine (in industrial strengths).

✓ **Key Point:** Because type of hypochlorite is not 100% pure chlorine, more pounds per day must be fed into the system to obtain the same amount of chlorine for disinfection. This is an important economical consideration for those facilities thinking about substituting hypochlorite for chlorine. Some studies indicate that such a substitution can increase operating costs, overall, by up to three times the cost of using chlorine.

To calculate the pounds per day hypochlorite required, a two-step calculation is required:

Step 1: $\text{mg/L}\ (\text{MGD})\ (8.34) = \text{lb/day}$

$$\text{Step 2: } \frac{\text{Chorine, lb/day}}{\frac{\%\text{ available}}{100}} = \text{hypochlorite, lb/day} \tag{8.6}$$

Example 8.7

Problem:
A total chlorine dosage of 10 mg/L is required to treat a particular wastewater. If the flow is 1.4 MGD and the hypochlorite has 65% available chlorine, how many pounds per day of hypochlorite will be required?

Solution:
Step 1: Calculate the pounds per day chlorine required using the milligrams per liter to pounds per day equation:

$$(\text{mg/L})\ (\text{MGD})\ (8.34) = \text{lb/day}$$
$$(10\ \text{mg/L})\ (1.4\ \text{MGD})\ (8.34\ \text{lb/gal}) = 117\ \text{lb/day}$$

Step 2: Calculate the pounds per day hypochlorite required. Because only 65% of the hypochlorite is chlorine, more than 117 lb/day will be required:

$$\frac{117\ \text{lb/day chlorine}}{\frac{65\ \text{available chlorine}}{100}} = 180\ \text{lb/day hypochlorite}$$

Example 8.8

Problem:
A wastewater flow of 840,000 gpd requires a chlorine dose of 20 mg/L. If sodium hypochlorite (15% available chlorine) is to be used, how many pounds per day of sodium hypochlorite are required? How many gallons per day of sodium hypochlorite is this?

Solution:

(1) Calculate the pounds per day chlorine required:

$$(\text{mg/L})\ (\text{MGD})\ (8.34) = \text{lb/day}$$
$$(20\ \text{mg/L})\ (0.84\ \text{MGD})\ (8.34\ \text{lb/gal}) = 140\ \text{lb/day chlorine}$$

(2) Calculate the pounds per day sodium hypochlorite:

$$\frac{140\ \text{lb/day chlorine}}{\frac{15\ \text{available chlorine}}{100}} = 933\ \text{lb/day hypochlorite}$$

(3) Calculate the gallons per day sodium hypochlorite:

$$\frac{933\ \text{lb/day}}{8.34\ \text{lb/gal}} = 112\ \text{gal/day sodium hypochlorite}$$

Example 8.9

Problem:
How many pounds of chlorine gas are necessary to treat 5,000,000 gal of wastewater at a dosage of 2 mg/L?

Solution:

Step 1: Calculate the pounds of chlorine required.

$$\text{V, } 10^6\text{gal} = \text{chlorine concentration } (\text{mg/L}) \times 8.34 = \text{lb chlorine}$$

Step 2: Substitute $5 \times 10^6\text{-gal} \times 2 \text{ mg/L} \times 8.34 = 83 \text{ lb chlorine}$

8.1.4 Chemical Solutions

A *water solution* is a homogeneous liquid consisting of the *solvent* (the substance that dissolves another substance) and the *solute* (the substance that dissolves in the solvent). Water is the solvent (see Figure 8.2). The solute (whatever it may be) may dissolve up to a certain point. This is called its *solubility*—that is, the solubility of the solute in the particular solvent (water) at a particular temperature and pressure. Remember, in chemical solutions, the substance being dissolved is called the *solute*, and the liquid present in the greatest amount in a solution (and that does the dissolving) is called the *solvent.* We should also be familiar with another term, *concentration*—the amount of solute dissolved in a given amount of solvent. Concentration is measured as:

$$\text{\% Strength} = \frac{\text{Wt. of solute}}{\text{Wt. of solution}} \times 100 = \frac{\text{Wt of solute}}{\text{Wt. of solute} + \text{solvent}} \times 100 \quad (8.7)$$

Example 8.10

Problem:
If 30 lb of chemical are added to 400 lb of water, what is the percent strength (by weight) of the solution?

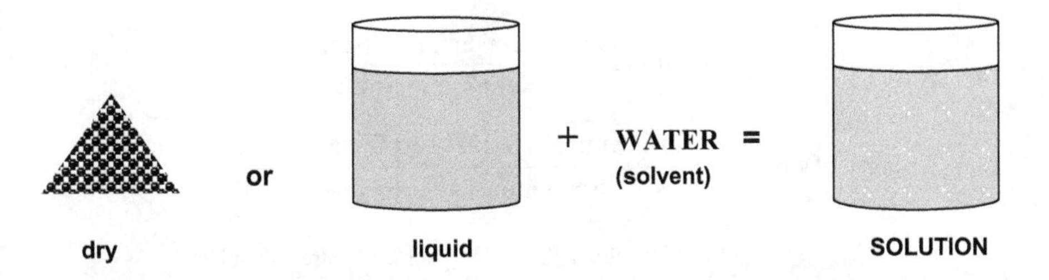

FIGURE 8.2 Components of a solution.

Solution:

$$\% \text{ Strength} = \frac{30 \text{ lb solute}}{400 \text{ lb water}} \times 100 = \frac{30 \text{ lb solute}}{30 \text{ lb solute} + 400 \text{ lb water}} \times 100$$

$$= \frac{30 \text{ lb solute}}{430 \text{ lb solute/water}} \times 100$$

$$\% \text{ Strength} = 7.0\%$$

It is important when making accurate computations of chemical strength to have a complete understanding of the dimensional units involved—for example, it is important to understand exactly what milligrams per liter (mg/L) signifies.

$$\text{Milligrams per Liter } (\text{mg/L}) = \frac{\text{Milligrams of Solute}}{\text{Liters of Solution}} \tag{8.8}$$

Another important dimensional unit commonly used when dealing with chemical solutions is *parts per million* (ppm).

$$\text{Parts per Million } (\text{ppm}) = \frac{\text{Parts of Solute}}{\text{Million Parts of Solution}} \tag{8.9}$$

✓ **Key Point:** "Parts" is usually a weight measurement, for example:

$$8 \text{ ppm} = \frac{8 \text{ lb solids}}{1{,}000{,}000 \text{ lb solution}}$$

$$8 \text{ ppm} = \frac{8 \text{ mg solids}}{1{,}000{,}000 \text{ mg solution}}$$

8.1.5 Mixing Solutions of Different Strength

When different percent strength solutions are mixed, we use the following equations, depending upon the complexity of the problem:

$$\% \text{ Strength of mixture} = \frac{\text{Chemical in Mixture, lb}}{\text{Solution Mixture, lb}} \times 100 \tag{8.10}$$

$$\% \text{ Strength of mixture} = \frac{\text{lb Chemical } (\text{Sol. 1}) + \text{lb Chem } (\text{Sol. 2})}{\text{lb Solution 1} + \text{lb Solution 2}} \times 100 \tag{8.11}$$

$$\% \text{ Strength of mixture} = \frac{\frac{(\text{Sol 1, lb})(\% \text{ Strength Sol 1})}{100} + \frac{(\text{Sol 2})(\% \text{ Strength Sol 2})}{100}}{\text{lb Solution 1} + \text{lb Solution 2}} \times 100 \tag{8.12}$$

Example 8.11

Problem:
If 25 lb of a 10% strength solution are mixed with 40 lb of 1% strength solutions, what is the percent strength of the solution mixture?

$$\% \text{ Strength of mixture} = \frac{\frac{(\text{Sol 1, lb})(\% \text{ Strength Sol 1})}{100} + \frac{(\text{Sol 2})(\% \text{ Strength Sol 2})}{100}}{\text{lb Solution 1} + \text{lb Solution 2}} \times 100$$

$$= \frac{(25 \text{ lb})\ (0.1) + (40 \text{ lb})\ (0.01)}{25 \text{ lb} + 40 \text{ lb}} \times 100$$

$$= \frac{2.5 \text{ lb} + 0.4 \text{ lb}}{65 \text{ lb}} \times 100$$

$$= 4.5\%$$

✓ **Key Point:** Percent strength should be expressed in terms of pounds chemical per pounds solution. That is, when solutions are expressed, for example, in terms of gallons, the gallons should be expressed as pounds before continuing with the percent strength calculations.

8.1.6 Solution Mixtures Target Percent Strength

When two different percent strength solutions are mixed in order to obtain a desired quantity of solution and a target percent strength, we use equation 8.12. After filling in the given information, we find for the unknown, x.

Example 8.12

Problem:

What weights of a 3% solution and a 6% solution must be mixed to make 800 lb of a 4% solution?

Solution:

$$\% \text{ Strength of mixture} = \frac{\frac{(\text{Sol 1, lb})(\% \text{ Strength Sol 1})}{100} + \frac{(\text{Sol 2})(\% \text{ Strength Sol 2})}{100}}{\text{lb Solution 1} + \text{lb Solution 2}} \times 100$$

$$4 = \frac{(x \text{ lb})(0.03) + (800 - x \text{ lb})\ (0.06)}{800 \text{ lb}} \times 100$$

$$\frac{(4)\ (800) = 0.03x + 48 - 0.06x}{100}$$

$$34 = \text{-}0.03x48$$

$$0.03x = 14$$

$$x = 467 \text{ lb of 3\% Solution}$$

$$\text{Then, } 800 - 467 = 333 \text{ lb of 6\% Solution}$$

8.1.7 Solution Chemical Feeder Setting (GPD)

Calculating gallons per day feeder setting depends on how the solution concentration is expressed, pounds per gallon or percent. If the solution strength is expressed as pounds per gallon, use the following equation:

$$\text{Solution, gpd} = \frac{(\text{Chemical, mg/L})\ (\text{Flow, MGD})\ (8.34, \text{lb/gal})}{\text{lb Chemical Solution}} \quad (8.13)$$

In water/wastewater operations, a standard, trial-and-error method known as jar testing is conducted to determine optimum chemical dosage. Jar testing has been the accepted bench testing procedure for many years. After jar testing results are analyzed to determine the best chemical dosage, the following example problems demonstrate how the actual calculations are made.

Example 8.13

Problem:
Jar tests indicate that the best liquid alum dose for water is 8 mg/L. The flow to be treated is 1.85 MGD. Determine the gallons per day setting for the liquid alum chemical feeder if the liquid alum contains 5.30 lb of alum per gallon of solution.

Solution:
First, calculate the pounds per day of dry alum required using the milligrams per liter to pounds per day equation:

$$\begin{aligned} \text{lb/day} &= (\text{dose, mg/L})\ (\text{flow, MGD})\ (8.34,\ \text{lb/gal}) \\ &= (8\ \text{mg/L})\ (1.85\ \text{MGD})\ (8.34\ \text{lb/gal}) \\ &= 123\ \text{lb/day dry alum} \end{aligned}$$

Then, calculate gallons per day solution required.

$$\text{Alum Solution, gpd} = \frac{123\ \text{lb/day alum}}{5.30\ \text{lb alum/gal solution}}$$

$$\text{Feeder Setting} = 23\ \text{gpd Alum Solution}$$

If the solution strength is expressed as a percent, we use the following equation:

$$\begin{aligned} &(\text{Chem., mg/L})\ (\text{Flow Treated, MGD})\ (8.34\ \text{lb/gal}) = \\ &(\text{Sol., mg/L})\ (\text{Sol. Flow, MGD})\ (8.34\ \text{lb/gal}) \end{aligned} \tag{8.14}$$

Example 8.14

Problem:
The flow to a plant is 3.40 MGD. Jar testing indicates that the optimum alum dose is 10 mg/L. What should the gallons per day setting be for the solution feeder if the alum solution is a 52% solution?

Solution:
A solution concentration of 52% is equivalent to 520,000 mg/L:

$$\text{Desired Dose, lb/day} = \text{Actual Dose, lb/day}$$

$$\begin{aligned} &(\text{Chemical, mg/L})\ (\text{Flow Treated, MGD}) = (\text{Sol, mg/L}) \\ &(8.34,\ \text{lb/gal}) \qquad\qquad (\text{Sol. Flow, MGD})\ (8.34\ \text{lb/gal} \end{aligned}$$

$$(10\text{ mg/L})\ (3.40\text{ MGD})\ (8.34\text{ lb/gal}) = (520{,}000\text{ mg/L})\ (x\text{MGD})\ (8.34\text{ lb/gal})$$

$$x = \frac{(10)\ (3.40)\ (8.34)}{(520{,}000)\ (8.34)}$$

$$x = 0.000065\text{ MGD}$$

This can be expressed as gallons per day flow:

$$0.000065\text{ MGD} = 65\text{ gpd flow}$$

8.1.8 Chemical Feed Pump—Percent Stroke Setting

Chemical feed pumps are generally positive displacement pumps (also called "piston" pumps). This type of pump displaces, or pushes out, a volume of chemical equal to the volume of the piston. The length of the piston, called the stroke, can be lengthened or shortened to increase or decrease the amount of chemical delivered by the pump. As mentioned, each stroke of a piston pump "displaces" or pushes out chemical. In calculating percent stroke setting, use the following equation:

$$\%\text{ Stroke Setting} = \frac{\text{Required Feed, gpd}}{\text{Maximum Feed, gpd}} \tag{8.15}$$

Example 8.15

Problem:
The required chemical pumping rate has been calculated as 8 gpm. If the maximum pumping rate is 90 gpm, what should the percent stroke setting be?

Solution:
The percent stroke setting is based on the ratio of the gallons per minute required to the total possible gallons per minute:

$$\%\text{ Stroke Setting} = \frac{\text{Required Feed, gpd}}{\text{Maximum Feed, gpd}}$$

$$= \frac{8\text{ gpm}}{90\text{ gpm}} \times 100$$

$$= 8.9\%$$

8.1.9 Chemical Solution Feeder Setting (mL/Min)

Some chemical solution feeders dispense chemical as milliliters per minute (mL/min). To calculate the milliliters per minute solution required, use the following equation:

$$\text{Solution, mL/min} = \frac{(\text{gpd})\ (3785\text{ mL/gal})}{1440\text{ min/day}} \tag{8.16}$$

Example 8.16

Problem:
The desired solution feed rate was calculated to be 7 gpd. What is this feed rate expressed as milliliters per minute?

Solution:
Because the gallons per day flow has already been determined, the milliliters per minute flow rate can be calculated directly:

$$\text{feed rate mL/min} = \frac{(\text{gpd})\ (3785\ \text{mL/gal})}{1440\ \text{min/day}}$$

$$= \frac{(7\ \text{gpd})\ (3785\ \text{mL/gal})}{1440\ \text{min/day}}$$

$$= 18\ \text{mL/min feed rate}$$

8.1.10 Chemical Feed Calibration

Routinely, to ensure accuracy, we need to compare the actual chemical feed rate with the feed rate indicated by the instrumentation. To accomplish this, we use calibration calculations. To calculate the actual chemical feed rate for a dry chemical feed, place a container under the feeder, weigh the container when empty, and then weigh the container again after a specified length of time, such as 30 min. Then, actual chemical feed rate can then be determined as:

$$\text{Chemical Feed Rate, lb/min} = \frac{\text{Chemical Applied, lb}}{\text{Length of Application, min}} \quad (8.17)$$

Example 8.17

Problem:
Calculate the actual chemical feed rate, in pounds per day, if a container is placed under a chemical feeder and a total of 2.2 lb is collected during a 30 min period.

Solution:
First calculate the pounds per minute feed rate:

$$\text{Chemical Feed Rate, lb/min} = \frac{\text{Chemical Applied, lb}}{\text{Length of Application, min}}$$

$$= \frac{2.2\ \text{lb}}{30\ \text{min}}$$

$$= 0.07\ \text{lb/min Feed Rate}$$

Then, calculate the pounds per day feed rate:

$$\text{Chemical Feed Rate, lb/day} = (0.07 \text{ lb/min})\ (1440 \text{ min/day})$$
$$= 101 \text{ lb/day Feed Rate}$$

Example 8.18

Problem:
A chemical feeder is to be calibrated. The container to be used to collect chemical is placed under the chemical feeder and weighed (0.35 lb). After 30 min, the weight of the container and chemical is found to be 2.2 lb. Based on this test, what is the actual chemical feed rate, in pounds per day?

Solution:
First, calculate the pounds per minute feed rate:

✓ **Key Point:** The chemical applied is the weight of the container and chemical minus the weight of the empty container.

$$\text{Chemical Feed Rate, lb/min} = \frac{\text{Chemical Applied, lb}}{\text{Length of Application, min}}$$
$$= \frac{2.2 \text{ lb} - 0.35 \text{ lb}}{30 \text{ minutes}}$$
$$= \frac{1.85 \text{ lb}}{30 \text{ minutes}}$$
$$= 0.062 \text{ lb/min Feed Rate}$$

Then, calculate the pounds per day feed rate:

$$(0.062 \text{ lb/min})\ (1440 \text{ min/day}) = 89 \text{ lb/day Feed Rate}$$

When the chemical feeder is for a solution, the calibration calculation is slightly more difficult than that for a dry chemical feeder. As with other calibration calculations, the actual chemical feed rate is determined and then compared with the feed rate indicated by the instrumentation. The calculations used for solution feeder calibration are as follows:

$$\text{Flow rate, gpd} = \frac{(\text{mL/min})\ (1440 \text{ min/day})}{3785 \text{ mL/gal}} = \text{gpd} \tag{8.18}$$

Then, calculate chemical dosage, in pounds per day:

$$\text{Chemical, lb/day} = (\text{Chemical, mg/L})\ (\text{Flow, MGD})\ (8.34\text{lb/day}) \tag{8.19}$$

Example 8.19

Problem:
A calibration test is conducted for a solution chemical feeder. For 5 min, the solution feeder delivers a total of 700 mL. The polymer solution is a 1.3% solution. What is the pounds per day feed rate? (Assume the polymer solution weighs 8.34 lb/gal.)

Problem:
The milliliters per minute flow rate is calculated as:

$$\frac{700 \text{ mL}}{5 \text{ min}} = 140 \text{ mL/min}$$

Then, convert milliliters per minute flow rate to gallons per day flow rate:

$$\frac{(140\ mL/min)(1440\ min/day)}{3785 \text{ mL/gal}} = 53 \text{ gpd flow rate}$$

And calculate pounds per day feed rate:

$$(\text{Chemical, mg/L})\ (\text{Flow, MGD})\ (8.34 \text{ lb/day}) = \text{Chemical, lb/day}$$

$$(13{,}000 \text{ mg/L})\ (0.000053 \text{ MGD})\ (8.34 \text{ lb/day}) = 5.7 \text{lbs/day polymer}$$

Actual pumping rates can be determined by calculating the volume pumped during a specified time frame. For example, if 120 gal are pumped during a 15 min test, the average pumping rate during the test is 8 gpm.

The gallons pumped can be determined by measuring the drop in tank level during the timed test.

$$\text{Flow, gpm} = \frac{\text{Volume Pumped, gal}}{\text{Duration of Test, min}} \tag{8.20}$$

Then, the actual flow rate (gpm) is calculated using:

$$\text{Flow, gpm} = \frac{(0.785)\ (D^2)\ (\text{Drop in Level, ft})\ (7.48 \text{ gal/cu ft})}{\text{Duration of Test, min}} \tag{8.21}$$

Example 8.20

Problem:
A pumping rate calibration test is conducted for a 5 min period. The liquid level in the 4 ft diameter solution tank is measured before and after the test. If the level drops 0.4 ft during the 5 min test, what is the pumping rate in gallons per minute?

Solution:

$$\text{Flow Rate, gpm} = \frac{(0.785)\ (D^2)\ (\text{Drop, ft})\ (7.48 \text{ gal/cu ft})}{\text{Duration of Test, min}}$$

$$= \frac{(0.785)\ (4 \text{ ft})\ (4 \text{ ft})\ (0.4 \text{ ft})\ (7.48 \text{ gal/cu ft})}{5 \text{ min}}$$

$$\text{Pumping Rate} = 38 \text{ gpm}$$

8.1.11 Average Use Calculations

During a typical shift, operators log in or record several parameter readings. The data collected is important in monitoring plant operation—in providing information on how to best optimize plant or unit process operation. One of the important parameters monitored each shift or each day is actual use of chemicals. From the recorded chemical use data, expected chemical use can be forecasted. This data also is important for inventory control, that is, determination can be made when additional chemical supplies will be required. To determine average chemical use, we first must determine the average chemical use:

$$\text{Average Use, lb/day} = \frac{\text{Total Chemical Used, lb}}{\text{Number of Days}} \tag{8.22}$$

or

$$\text{Average Use, gpd} = \frac{\text{Total Chemical Used, gal}}{\text{Number of Days}} \tag{8.23}$$

Then calculate day's supply in inventory:

$$\text{Day's Supply in Inventory} = \frac{\text{Total Chemical in Inventory, lb}}{\text{Average Use, lb/day}} \tag{8.24}$$

or

$$\text{Day's Supply in Inventory} = \frac{\text{Total Chemical in Inventory, gal}}{\text{Average Use, gpd}} \tag{8.25}$$

Example 8.21

Problem:
The chemical used for each day during a week is given in the following. Based on this data, what was the average pounds per day chemical use during the week?

Monday—92 lb/day
Tuesday—94 lb/day
Wednesday—92 lb/day
Thursday—88 lb/day
Friday—96 lb/day
Saturday—92 lb/day
Sunday—88 lb/day

Solution:

$$\text{Average Use, lb/day} = \frac{\text{Total Chemical Used, lb}}{\text{Number of Days}}$$

$$= \frac{642\text{ lb}}{7\text{ days}}$$

$$\text{Average Use} = 91.7\text{ lb/day}$$

Example 8.22

Problems:
The average chemical use at a plant is 83 lb/day. If the chemical inventory in stock is 2,600 lb, how many days' supply is this?

Source:

$$\text{Days' Supply in Inventory} = \frac{\text{Total Chemical in Inventory, lb}}{\text{Average Use, lb/day}}$$
$$= \frac{\text{2600 lb in Inventory}}{\text{83 lb/day Average Use}}$$
$$= \text{31.3 days' supply}$$

8.2 CHEMICAL DOSAGE PRACTICE PROBLEMS

Problem 8.1

Determine the chlorinator setting (pounds per day) needed to treat a flow of 8 MGD with a chemical dose of 5 mg/L.

Solution:

$$\text{Chemical, lb/day} = \text{Chemical, mg/L} \times \text{flow, MGD} \times \text{8.34 lb/gal}$$
$$= \text{5 mg/L} \times \text{8 MGD} \times \text{8.34 lb/gal}$$
$$= \text{333.6 lb/day}$$

Problem 8.2

The desired dosage for a dry polymer is 8 mg/L. If the flow to be treated is 2,000,000 gpd, how many pounds per day polymer will be required?

Solution:

$$\text{Polymer, lb/day} = \text{Polymer, mg/L} \times \text{flow, MGD} \times \text{8.34 lb/day}$$
$$= \text{8 mg/L Polymer} \times (\text{2.0 MGD})\ (\text{8.34 lb/day})$$
$$= \text{133.4 lb/day Polymer}$$

Problem 8.3

To neutralize a sour digester, 1 lb of lime is added for every pound of volatile acids in the digester biosolids. If the digester contains 300,000 gal of biosolids with a volatile acid (VA) level of 2,000 mg/L, how many pounds of lime should be added?

Solution:
Because volatile acid concentration is 2,000 mg/L, the lime concentration should also be 2,000 mg/L:

$$\begin{aligned} \text{lb lime required} &= \text{lime, mg/L} \times \text{digester volume, MG} \times 8.34 \text{ lb/gal} \\ &= (2000 \text{ mg/L})\ (0.30 \text{ MG})\ (8.34 \text{ lb/gal}) \\ &= 5{,}004 \text{ lb lime} \end{aligned}$$

Problem 8.4

Determine the chlorinator setting (pounds per day) needed to treat a flow of 9 MGD with a chlorine dose of 5 mg/L.

Solution:

$$(\text{mg/L})(\text{MGD})(8.34) = \text{lb/day}$$

$$(5 \text{ mg/L})\ (9 \text{ MGD})\ (8.34 \text{ lb/gal} = 375 \text{ lb/day}$$

Problem 8.5

The chlorine dosage for a secondary effluent is 5 mg/L. If the chlorine residual after 30 min contact time is found to be 0.4 mg/L, what is the chlorine demand expressed in milligrams per liter?

Solution:

$$\text{Chlorine Dose} = \text{Chlorine demand} + \text{Chlorine Residual}$$

$$5 \text{ mg/L} = x \text{ mg/L} + 0.4 \text{ mg/L}$$

$$5 \text{ mg/L} - 0.4 \text{ mg/L} = x \text{ mg/L}$$

$$x = 4.6 \text{ mg/L Chlorine Demand}$$

9 Biosolids Production and Pumping Calculations

9.1 PROCESS RESIDUALS

Wastewater unit treatment processes remove solids and biochemical oxygen demand (BOD) from the waste stream before the liquid effluent is discharged to its receiving waters. What remains to be disposed of is a mixture of solids and wastes, called *process residuals*—more commonly referred to as biosolids (or sludge).

✓ **Key Point:** *Sludge* is the commonly accepted name for wastewater residual solids. However, if wastewater sludge is used for beneficial reuse (i.e., as a soil amendment or fertilizer), it is commonly called biosolids. We choose to refer to *process residuals* as *biosolids* in this text.

The most costly and complex aspect of wastewater treatment can be the collection, processing, and disposal of biosolids. This is the case because the quantity of biosolids produced may be as high as 2% of the original volume of wastewater, depending somewhat on the treatment process being used. Because the 2% biosolids can be as much as 97% water content, and because cost of disposal will be related to the volume of biosolids being processed, one of the primary purposes or goals (along with stabilizing it so it is no longer objectionable or environmentally damaging) of biosolids treatment is to separate as much of the water from the solids as possible.

9.2 PRIMARY AND SECONDARY SOLIDS PRODUCTION CALCULATIONS

It is important to point out that when making calculations pertaining to solids and biosolids, the term "solids" refers to dry solids, and the term "biosolids" refers to the solids and water. The solids produced during primary treatment depend on the solids that settle in, or are removed by, the primary clarifier (see Figure 9.1). In making primary clarifier solids production calculations, we use the milligrams per liter to pounds per day equation, shown next.

$$\text{Suspended Solids (SS) Removed, lb/day} = (\text{SS Removed, mg/L})(\text{Flow, MGD})(8.34\text{ lb/gal}) \tag{9.1}$$

DOI: 10.1201/9781003354314-9

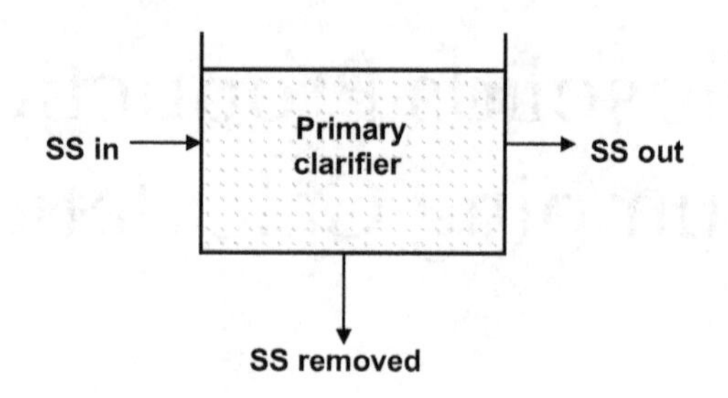

FIGURE 9.1 Solids in and out of a primary clarifier.

9.2.1 Primary Clarifier Solids Production Calculations

Example 9.1

Problem:
A primary clarifier receives a flow of 1.80 MGD with suspended solids concentrations of 340 mg/L. If the clarifier effluent has a suspended solids concentration of 180 mg/L, how many pounds of solids are generated daily?

Solution:

$$\text{SS, lbs/day Removed} = (\text{SS Removed, mg/L})\ (\text{Flow, MGD})\ (8.34\ \text{lb/gal})$$
$$= (160\ \text{mg/L})\ (1.80\ \text{MGD})\ (8.34\ \text{lb/gal})$$
$$\text{Solids} = 2402\ \text{lb/day}$$

Example 9.2

Problem:
The suspended solids content of the primary influent is 350 mg/L, and the primary influent is 202 mg/L. How many pounds of solids are produced during a day that the flow is 4,150,000 gpd?

Solution:

$$\text{SS, lbs/day Removed} = (\text{SS Removed, mg/L})\ (\text{Flow, MGD})\ (8.34\ \text{lb/gal})$$
$$= (148\ \text{mg/L})\ (4.15\ \text{MGD})\ (8.34\ \text{lb/gal})$$
$$\text{Solids Removed} = 5122\ \text{lb/day}$$

9.2.2 Secondary Clarifier Solids Production Calculation

Solids produced during secondary treatment depend on many factors, including the amount of organic matter removed by the system and the growth rate of the bacteria (see Figure 9.2). Because precise calculations of biosolids production are complex, we use a rough estimate method of solids production which uses an estimated growth

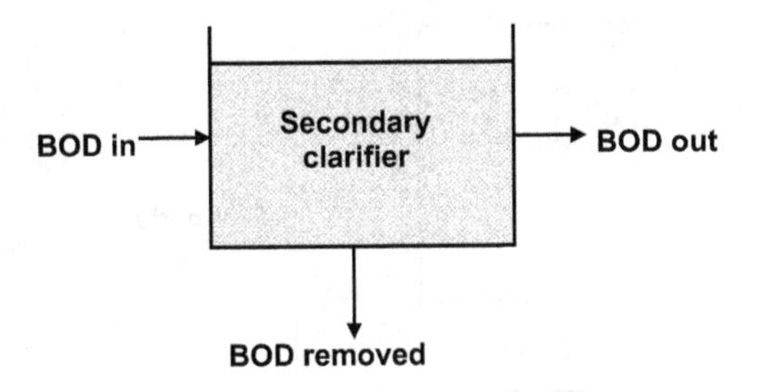

FIGURE 9.2 Solids produced during secondary treatment.

rate (unknown) value. We use the BOD removed, pounds per day equation, shown here:

$$\text{BOD Removed, lb/day} = (\text{BOD Removed, mg/L})(\text{Flow, MGD})(8.34\ \text{lb/day}) \tag{9.2}$$

Example 9.3

Problem:
The 1.5 MGD influent to the secondary system has a BOD concentration of 174 mg/L. The secondary effluent contains 22 mg/L BOD. If the bacteria growth rate, unknown *x-value*, for this plant is 0.44 lb SS/lb BOD removed, how many pounds of dry biosolids solids are produced each day by the secondary system?

Solution:

$$\begin{aligned}\text{BOD Removed, lb/day} &= (\text{BOD, mg/L})(\text{Flow, MGD})(8.34\ \text{lb/gal})\\ &= (152\ \text{mg/L})(1.5\ \text{MGD})\ 8.34\ \text{lb/gal}\\ &= 1902\ \text{lb/day}\end{aligned}$$

Then use the unknown *x-value* to determine pounds per day solids produced.

$$\frac{0.44\ \text{lb SS Produced}}{1\ \text{lb BOD Removed}} = \frac{x\ \text{lb SS Produced}}{1902\ \text{lb/day BOD Removed}}$$

$$\frac{(0.44)(1902)}{1} = x$$

$$837\ \text{lb/day Solids Produced} = x$$

✓ **Key Point:** Typically, for every pound of food consumed (BOD removed) by the bacteria, between 0.3 and 0.7 lb of new bacteria cells are produced; these are solids that have to be removed from the system.

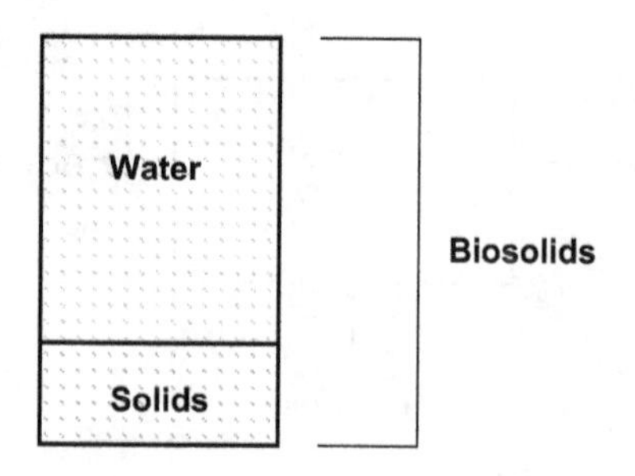

FIGURE 9.3 Solids need to be removed.

9.3 PERCENT SOLIDS

Biosolids are composed of water and solids. The vast majority of biosolids is water—usually in the range of 93–97% (see Figure 9.3). To determine the solids content of a biosolids, a sample of biosolids is dried overnight in an oven at 103°–105°F. The solids that remain after drying represent the total solids content of the biosolids. Solids content may be expressed as a percent or as milligrams per liter. Either of two equations is used to calculate percent solids:

$$\%\text{ Solids} = \frac{\text{Total Solids, g}}{\text{Biosolids Sample, g}} \times 100 \tag{9.3}$$

$$\%\text{ Solids} = \frac{\text{Solids, lb/day}}{\text{Biosolids, lb/day}} \times 100 \tag{9.4}$$

Example 9.4

Problem:
The total weight of a biosolids sample (sample only, not the dish) is 22 g. If the weight of the solids after drying is 0.77 g, what is the percent total solids of the biosolids?

Solution:

$$\begin{aligned}\%\text{ Solids} &= \frac{\text{Total Solids, grams}}{\text{Biosolids Sample, grams}} \times 100 \\ &= \frac{0.77\text{ grams}}{22\text{ grams}} \times 100 \\ &= 3.5\%\end{aligned}$$

9.4 BIOSOLIDS PUMPING

While on shift, wastewater operators are often required to make various process control calculations. An important calculation involves biosolids pumping. The biosolids pumping calculations the operator may be required to make are covered in this section.

9.4.1 Estimating Daily Biosolids Production

The calculation for estimation of the required biosolids pumping rate provides a method to establish an initial pumping rate or to evaluate the adequacy of the current withdrawal rate.

$$\text{Est. pump rate} = \frac{(\text{Influ. TSS Conc.} - \text{Effluent TSS Conc.}) \times \text{Flow} \times 8.34}{\%\ \text{Solids in Sludge} \times 8.34 \times 1{,}440\ \text{min/day}} \tag{9.5}$$

Example 9.5

Problem:
The biosolids withdrawn from the primary settling tank contains 1.4% solids. The unit influent contains 285 mg/L TSS, and the effluent contains 140 mg/L TSS. If the influent flow rate is 5.55 MGD, what is the estimated biosolids withdrawal rate in gallons per minute (assuming the pump operates continuously)?

Solution:

$$\text{Biosolids Rate, gpm} = \frac{(285\ \text{mg/L} - 140\ \text{mg/L}) \times 5.55 \times 8.34}{0.014 \times 8.34 \times 1{,}440\ \text{min/day}} = 40\ \text{gpm}$$

9.4.2 Biosolids Production in Pounds/Million Gallons

A common method of expressing biosolids production is in pounds of biosolids per million gallons of wastewater treated.

$$\text{Biosolids, lb/MG} = \frac{\text{Total Biosolids Production, lb}}{\text{Total Wastewater Flow, MG}} \tag{9.6}$$

Example 9.6

Problem:
Records show that the plant has produced 85,000 gal of biosolids during the past 30 days. The average daily flow for this period was 1.2 MGD. What was the plant's biosolids production in pounds per million gallons?

Solution:

$$\text{Biosolids, lb/MG} = \frac{85{,}000\text{-gal} \times 8.34\ \text{lb/gal}}{1.2\ \text{MGD} \times 30\ \text{days}} = 19{,}692\ \text{lb/MG}$$

9.4.3 Biosolids Production in Wet Tons/Year

Biosolids production can also be expressed in terms of the amount of biosolids (water and solids) produced per year. This is normally expressed in wet tons per year.

$$\text{Biosolids, Wet Tons/year} = \frac{\text{Biosolids Prod., lb/MG} \times \text{Ave. Daily Flow, MGD} \times 365 \text{ days/year}}{2{,}000 \text{ lb/ton}} \tag{9.7}$$

Example 9.7

Problem:
The plant is currently producing biosolids at the rate of 16,500 lb/MG. The current average daily wastewater flow rate is 1.5 MGD. What will be the total amount of biosolids produced per year in wet tons per year?

Solution:

$$\text{Biosolids, Wet Tons/year} = \frac{16{,}500 \text{ lb/MG} \times 1.5 \text{ MGD} \times 365 \text{ days/year}}{2{,}000 \text{ lb/ton}}$$

$$= 4{,}517 \text{ Wet Tons/year}$$

9.4.4 Biosolids Pumping Time

The biosolids pumping time is the total time the pump operates during a 24 hr period in minutes.

$$\text{Pump Operating Time} = \text{Time/Cycle, minutes} \times \text{Frequency, cycles/day} \tag{9.8}$$

- Note: The following information is used for examples 9.8–9.13.

Frequency	24 times/day
Pump rate	120 gpm
Solids	3.70%
Volatile matter	66%

Example 9.8

Problem:
What is the pump operating time?

Solution:

$$\text{Pump Operating Time} = 15 \text{ min/hour} \times 24 \text{ (cycles)/day} = 360 \text{ min/day}$$

9.4.5 Biosolids Pumped/Day in Gallons

$$\text{Biosolids, gpd} = \text{Operating Time, min/day} \times \text{Pump Rate, gpm} \tag{9.9}$$

Example 9.9

Problem:
What is the biosolids pumped per day in gallons?

Solution:

$$\text{Biosolids, gpd} = 360 \text{ min/day} \times 120 \text{ gpm} = 43{,}200 \text{ gpd}$$

9.4.6 Biosolids Pumped/Day in Pounds

$$\text{Sludge, lb/day} = \text{Gallons of Biosolids Pumped} \times 8.34 \text{ lb/gal} \tag{9.10}$$

Example 9.10

Problem:
What is the biosolids pumped per day in pounds?

Solution:

$$\text{Biosolids, lb/day} = 43{,}200 \text{ gal/day} \times 8.34 \text{ lb/gal} = 360{,}000 \text{ lb/day}$$

9.4.7 Solids Pumped/Day in Pounds

$$\text{Solids Pumped, lb/day} = \text{Biosolids Pumped, lb/day} \times \% \text{ Solids}$$

Example 9.11

Problem:
What are the solids pumped per day?

Solution:

$$\text{Solids Pumped lb/day} = 360{,}300 \text{ lb/day} \times 0.0370 = 13{,}331 \text{ lb/day}$$

9.4.8 Volatile Matter Pumped/Day in Pounds

$$\text{Vol. Matter} \left(\text{lb/day}\right) = \text{Solids Pumped, lb/day} \times \% \text{ Volatile Matter} \tag{9.11}$$

Example 9.12

Problem:
What is the volatile matter in pounds per day?

Solution:

$$\text{Volatile Matter, lb/day} = 13{,}331 \text{ lb/day} \times 0.66 = 8{,}798 \text{ lb/day}$$

9.4.9 Pounds of Solids/Pounds of Volatile Solids/Day

If we wish to calculate the pounds of solids or the pounds of volatile solids removed per day, the individual equations demonstrated earlier can be combined into a single calculation.

$$\text{Solids, lb/day} = \text{Pump Time, min/cyc} \times \text{Freq., cyc/day} \times \text{Rage, gpm} \times 8.34\ \text{lb/gal} \times \text{solids} \tag{9.12}$$

$$\text{Vol. Mat., lb/day} = \text{Time, min/cyc} \times \text{Freq. Cyc/day} \times \text{Rate, gpm} \times 8.34 \times \%\ \text{Solids} \times \%\ \text{VM}$$

Example 9.13

Problem:
What are the solids and volatile solids removed per day?

Solution:

$$\text{Solids, lb/day} = 15\ \text{min/cyc} \times 24\ \text{cyc/day} \times 120\ \text{gpm} \times 8.34 \times 0.0370$$
$$= 13{,}331\ \text{lb/day}$$

$$\text{VM, lb/day} = 15\ \text{min/cyc} \times 24\ \text{cyc/day} \times 120\ \text{gpm} \times 8.34 \times 0.0370 \times .66$$
$$= 8{,}798\ \text{lb/day}$$

9.5 BIOSOLIDS PRODUCTION AND PUMPING PRACTICE CALCULATIONS

Problem 9.1

A primary clarifier receives a flow of 1.60 MGD with suspended solids concentrations of 360 mg/L. If the clarifier effluent has a suspended solids concentration of 180 mg/L, how many pounds of solids are generated daily?

Solution:

$$\text{SS, lb/day Removed} = (\text{SS Removed, mg/L})\ (\text{Flow, MGD})\ (8.34\ \text{lb/gal})$$
$$= (180\ \text{mg/L})\ (1.60\ \text{MGD})\ (8.34\ \text{lb/gal})$$
$$\text{Solids} = 2402\ \text{lb/day}$$

Problem 9.2

The suspended solids content of the primary influent is 300 mg/L, and the primary influent is 200 mg/L. How many pounds of solids are produced during a day that the flow is 4,100,000 gpd?

Solution:

$$\text{SS, lb/day Removed} = (\text{SS Removed, mg/l})\ (\text{Flow, MGD})\ (8.34\ \text{lb/gal})$$
$$= (100\ \text{mg/L})\ (4.1\ \text{MGD})\ (8.34\ \text{lb/gal})$$
$$\text{Solids Removed} = 3419\ \text{lb/day}$$

Problem 9.3

The 1.5 MGD influent to the secondary system has a BOD concentration of 170 mg/L. The secondary effluent contains 20 mg/L BOD. If the bacteria growth rate, unknown *x-value*, for this plant is 0.40 lb SS/lb BOD removed, how many pounds of dry biosolids solids are removed each day by the secondary system?

Solution:

$$\text{BOD Removed, lb/day} = (\text{BOD, mg/L})\ (\text{Flow, MGD})\ (8.34\ \text{lb/gal})$$
$$= (150\ \text{mg/L})\ (1.5\ \text{MGD})\ 8.34\ \text{lb/gal}$$
$$= 1877\ \text{lb/day BOD Removed}$$

Problem 9.4

The total weight of a biosolids sample (sample only, not the dish) is 22 g. If the weight of the solids after drying is 0.77 g, what is the percent total solids of the biosolids?

Solution:

$$\%\ \text{Solids} = \frac{\text{Total Solids (g)}}{\text{Biosolids Sample (g)}} \times 100$$
$$= \frac{0.77\ \text{g}}{22\ \text{g}} \times 100$$
$$= 3.5\%$$

Problem 9.5

The biosolids withdrawn from the primary settling tank contains 1.4% solids. The unit influent contains 280 mg/L TSS, and the effluent contains 150 mg/L TSS. If the influent flow rate is 5.0 MGD, what is the estimated biosolids withdrawal rate in gallons per minute (assuming the pump operates continuously)?

Solution:

$$\text{Biosolids Rate, gpm} = \frac{(280\ \text{mg/L} - 150\ \text{mg/L}) \times 5.0 \times 8.34}{0.014 \times 8.34 \times 1{,}440\ \text{min/day}} = 32\ \text{gpm}$$

10 Biosolids Thickening Calculations

10.1 THICKENING

Biosolids thickening (or concentration) is a unit process used to increase the solids content of the biosolids by removing a portion of the liquid fraction. In other words, biosolids thickening is all about volume reduction. By increasing the solids content, more economical treatment of the biosolids can be effected. Biosolids thickening processes include the following:

- Gravity thickeners
- Flotation thickeners
- Solids concentrators

Biosolids thickening calculations are based on the concept that the solids in the primary or secondary biosolids are equal to the solids in the thickened biosolids. The solids are the same (Figure 10.1). It is primarily water that has been removed in order to thicken the biosolids and result in higher percent solids. In this unthickened biosolids, the solids might represent 1% or 4% of the total pounds of biosolids. But when some of the water is removed, those same amount solids might represent 5–7% of the total pounds of biosolids.

✓ **Key Point:** The key to biosolids thickening calculations—solids remain constant (see Figure 10.1).

10.2 GRAVITY/DISSOLVED AIR FLOTATION THICKENER CALCULATIONS

As mentioned, biosolids thickening calculations are based on the concept that the solids in the primary or secondary biosolids are equal to the solids in the thickened biosolids. That is, assuming a negligible amount of solids is lost in the thickener overflow, the solids are the same. Note that the water is removed to thicken the biosolids and results in higher percent solids.

10.2.1 Estimating Daily Biosolids Production

The calculation for estimation of the required biosolids pumping rate provides a method to establish an initial pumping rate or to evaluate the adequacy of the current pump rate.

DOI: 10.1201/9781003354314-10

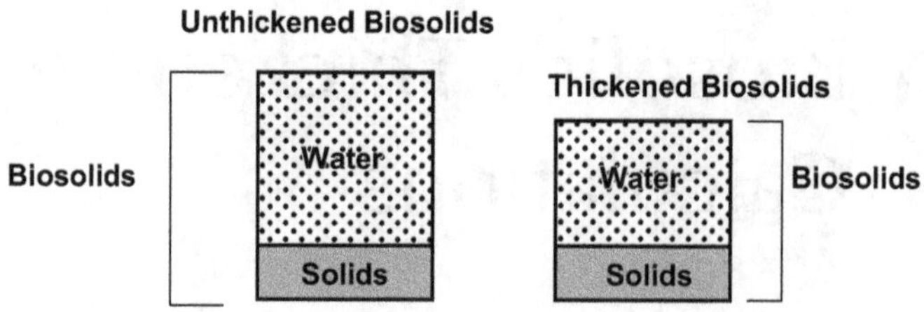

FIGURE 10.1 Biosolids thickening = volume reduction.

$$\text{Est. Pump Rate} = \frac{(\text{Influent TSS Conc.} - \text{Eff. TSS Conc.}) \times \text{Flow} \times 8.34}{\%\ \text{Solids in Biosolids} \times 8.34 \times 1{,}440\ \text{min/day}} \quad (10.1)$$

Example 10.1

Problem:
The biosolids withdrawn from the primary settling tank contains 1.5% solids. The unit influent contains 280 mg/L TSS, and the effluent contains 141 mg/L TSS. If the influent flow rate is 5.55 MGD, what is the estimated biosolids withdrawal rate in gallons per minute (assuming the pump operates continuously)?

Solution:

$$\text{Biosolids Withdrawal Rate, gpm} = \frac{(280\ \text{mg/L} - 141\ \text{mg/L}) \times 5.55\ \text{MGD} \times 8.34}{0.015 \times 8.34 \times 1{,}440\ \text{min/day}}$$

$$= 36\ \text{gpm}$$

10.2.2 Surface Loading Rate, gal/day/ft^2

Surface loading rate (surface settling rate) is hydraulic loading—the amount of biosolids applied per square foot of gravity thickener.

$$\text{Surface Loading, gal/day/ft}^2 = \frac{\text{Biosolids Applied to the Thickener, gpd}}{\text{Thickener Area, ft}^2} \quad (10.2)$$

Example 10.2

Problem:
A 70 ft diameter gravity thickener receives 32,000 gpd of biosolids. What is the surface loading in gallons per square foot per day?

Solution:

$$\text{Surface Loading} = \frac{32{,}000\ \text{gpd}}{0.785 \times 70\ \text{ft} \times 70\ \text{ft}} = 8.32\ \text{gpd/ft}^2$$

10.2.3 Solids Loading Rate, lb/day/ft²

The solids loading rate is the pounds of solids per day being applied to 1 sq ft of tank surface area. The calculation uses the surface area of the bottom of the tank. It assumes the floor of the tank is flat and has the same dimensions as the surface.

$$\text{Surface. Loading Rate, lb/day/ft}^2 = \frac{\text{\% Biosolids Solids} \times \text{Biosolids Flow, gpd} \times 8.34\ \text{lb/gal}}{\text{Thickener Area, ft}^2} \tag{10.3}$$

Example 10.3

Problem:
The thickener influent contains 1.6% solids. The influent flow rate is 39,000 gpd. The thickener is 50 ft in diameter and 10 ft deep. What is the solid loading in pounds per day?

Solution:

$$\text{Solids Loading Rate, lb/day/ft}^2 = \frac{0.016 \times 39{,}000\ \text{gpd} \times 8.34\ \text{lb/gal}}{0.785 \times 50\ \text{ft} \times 50\ \text{ft}} = 2.7\ \text{lb/ft}^2$$

10.2.4 Concentration Factor (CF)

The concentration factor (CF) represents the increase in concentration resulting from the thickener—it is a means of determining the effectiveness of the gravity thickening process.

$$\text{CF} = \frac{\text{Thickened Biosolids Concentration, \%}}{\text{Influent Biosolids Concentration, \%}} \tag{10.4}$$

Example 10.4

Problem:
The influent biosolids contains 3.5% solids. The thickened biosolids solids concentration is 7.7%. What is the concentration factor?

Solution:

$$\text{CF} = \frac{7.7\%}{3.5\%} = 2.2$$

Air-to-Solids Ratio
This is the ratio between the pounds of solids entering the thickener and the pounds of air being applied.

$$\text{Air:Solids Ratio} = \frac{\text{Air Flow ft}^3/\text{min} \times 0.0785\ \text{lb/ft}^3}{\text{Biosolids Flow, gpm} \times \text{\% Solids} \times 8.34\ \text{lb/gal}} \tag{10.5}$$

Example 10.5

Problem:
The biosolids pumped to the thickener is 0.85% solids. The airflow is 13 cfm. What is the air-to-solids ratio if the current biosolids flow rate entering the unit is 50 gpm?

Solution:

$$\text{Air:Solids Ratio} = \frac{13 \text{ cfm} \times 0.075 \text{ lb/ft}^3}{50 \text{ gpm} \times 0.0085 \times 8.34 \text{ lb/gal}} = 0.28$$

10.2.5 Recycle Flow in Percent

The amount of recycle flow expressed as a percent.

$$\text{Recycle \%} = \frac{\text{Recycle Flow Rate, gpm} \times 100}{\text{Sludge Flow, gpm}} = 175\% \tag{10.6}$$

Example 10.6

Problem:
The sludge flow to the thickener is 80 gpm. The recycle flow rate is 140 gpm. What is the percent recycle?

Solution:

$$\text{\% Recycle} = \frac{140 \text{ gpm} \times 100}{80 \text{ gpm}} = 175\%$$

10.3 CENTRIFUGE THICKENING CALCULATIONS

A centrifuge exerts a force on the biosolids thousands of times greater than gravity. Sometimes polymer is added to the influent of the centrifuge to help thicken the solids. The two most important factors that affect the centrifuge are the volume of the biosolids put into the unit (gpm) and the pounds of solids put in. The water that is removed is called centrate. Normally, hydraulic loading is measured as flow rate per unit of area. However, because of the variety of sizes and designs, hydraulic loading to centrifuges does not include area considerations. It is expressed only as gallons per hour. The equations to be used if the flow rate to the centrifuge is given as gallons per day or gallons per minute are:

$$\text{Hydraulic Loading, gph} = \frac{\text{Flow, gpd}}{24 \text{ hr/day}} \tag{10.7}$$

$$\text{Hydraulic Loading, gph} = (\text{gpm flow})\ (60 \text{ min/hr}) \tag{10.8}$$

Example 10.7

Problem:
A centrifuge receives a waste activated biosolids flow of 40 gpm. What is the hydraulic loading on the unit in gallons per hour?

Solution:

$$\text{Hydraulic Loading, gph} = (\text{gpm flow})\ (60\ \text{min/hr})$$
$$= (40\ gpm)(60\ min/hr)$$
$$= 2400\ \text{gph}$$

Example 10.8

Problem:
A centrifuge receives 48,600 gal of biosolids daily. The biosolids concentration before thickening is 0.9%. How many pounds of solids are received each day?

Solution:

$$\frac{48{,}600\ \text{gal}}{\text{d}} \times \frac{8.34\ \text{lb}}{\text{gal}} \times \frac{0.9}{100} = 3648\ \text{lb/d}$$

10.4 BIOSOLIDS THICKENING PRACTICE PROBLEMS

Problem 10.1

The biosolids withdrawn from the primary settling tank contains 1.5% solids. The unit influent contains 260 mg/L TSS, and the effluent contains 130 mg/L TSS. If the influent flow rate is 5 MGD, what is the estimated biosolids withdrawal rate in gallons per minute (assuming the pump operates continuously)?

Solution:

$$\text{Biosolids Withdrawal Rate, gpm} = \frac{(260\ \text{mg/L} - 130\ \text{mg/L}) \times 5.0\ \text{MGD} \times 8.34}{0.015 \times 8.34 \times 1{,}440\ \text{min/day}}$$
$$= 30\ \text{gpm}$$

Problem 10.2

A 60 ft diameter gravity thickener receives 31,000 gpd of biosolids. What is the surface loading in gallons per square foot per day?

Solution:

$$\text{Surface Loading} = \frac{31{,}000\ \text{gpd}}{0.785 \times 60\ \text{ft} \times 60\ \text{ft}} = 10.97\ \text{gpd/ft}^2$$

Problem 10.3

The thickener influent contains 1.6% solids. The influent flow rate is 38,000 gpd. The thickener is 55 ft in diameter and 11 ft deep. What is the solid loading in pounds per day?

Solution:

$$\text{Solids Loading Rate, lb/day/ft}^2 = \frac{0.016 \times 38{,}000 \text{ gpd} \times 8.34 \text{ lb/gal}}{0.785 \times 55 \text{ ft} \times 55 \text{ ft}} = 2.14 \text{ lb/ft}^2$$

Problem 10.4

The influent biosolids contain 3.3% solids. The thickened biosolids solids concentration is 7.5%. What is the concentration factor?

Solution:

$$\text{CF} = \frac{7.5\%}{3.3\%} = 2.27$$

Problem 10.5

The biosolids pumped to the thickener is 0.85% solids. The airflow is 15 cfm. What is the air-to-solids ratio if the current biosolids flow rate entering the unit is 60 gpm?

Solution:

$$\text{Air:Solids Ratio} = \frac{15 \text{ cfm} \times 0.075 \text{ lb/ft}^3}{60 \text{ gpm} \times 0.0085 \times 8.34 \text{ lb/gal}} = 0.26$$

11 Biosolids Digestion

11.1 BIOSOLIDS STABILIZATION

A major problem in designing wastewater treatment plants is the disposal of biosolids into the environment without causing damage or nuisance. Untreated biosolids is even more difficult to dispose of. Untreated raw biosolids must be stabilized to minimize disposal problems. In many cases, the term *stabilization* is considered synonymous with *digestion*.

✓ **Key Point:** The ***stabilization*** of organic matter is accomplished biologically using a variety of organisms. The microorganisms convert the colloidal and dissolved organic matter into various gases and into protoplasm. Because protoplasm has a specific gravity slightly higher than that of water, it can be removed from the treated liquid by gravity.

Biosolids digestion is a process in which biochemical decomposition of the organic solids occurs; in the decomposition process, the organics are converted into simpler and more stable substances. Digestion also reduces the total mass or weight of biosolids solids, destroys pathogens, and makes it easier to dry or dewater the biosolids. Well-digested biosolids have the appearance and characteristics of a rich potting soil. Biosolids may be digested under aerobic or anaerobic conditions. Most large municipal wastewater treatment plants use anaerobic digestion. Aerobic digestion finds application primarily in small package-activated biosolids treatment systems.

11.2 AEROBIC DIGESTION PROCESS CONTROL CALCULATIONS

The purpose of *aerobic digestion* is to stabilize organic matter, to reduce volume, and to eliminate pathogenic organisms. Aerobic digestion is similar to the activated biosolids process. Biosolids are aerated for 20 days or more. Volatile solids are reduced by biological activity.

11.2.1 VOLATILE SOLIDS LOADING IN LB/FT³/DAY

Volatile solids (organic matter) loading for the aerobic digester is expressed in pounds of volatile solids entering the digester per day per cubic foot of digester capacity.

$$\text{Volatile Solids Loading, lb/day/ft}^3 = \frac{\text{Volatile Solids Added, lb/day}}{\text{Digester Volume, ft}^3} \qquad (11.1)$$

DOI: 10.1201/9781003354314-11

Example 11.1

Problem:
The aerobic digester is 20 ft in diameter and has an operating depth of 20 ft. The biosolids that are added to the digester daily contains 1,500 lb of volatile solids. What is the volatile solids loading in pounds per day per cubic foot?

Solution:

$$\text{Vol. Solids Loading, lb/day/ft}^3 = \frac{1{,}500\text{ lb/day}}{0.785 \times 20\text{ ft} \times 20\text{ ft} \times 20\text{ ft}} = 0.24\text{ lb/day/ft}^3$$

11.2.2 Digestion Time (Day)

The theoretical time the biosolids remain in the aerobic digester:

$$\text{Digestion Time, Days} = \frac{\text{Digester volume, gallons}}{\text{Biosolids Added, gpd}} \tag{11.2}$$

Example 11.2

Problem:
The digester volume is 240,000 gal. Biosolids are added to the digester at the rate of 15,000 gpd. What is the digestion time in days?

Solution:

$$\text{Digestion Time, Days} = \frac{240{,}000\text{ gal}}{15{,}000\text{ gpd}} = 16\text{ days}$$

11.2.3 pH Adjustment

In many instances, the pH of the aerobic digester will fall below the levels required for good biological activity. When this occurs, the operator must perform a laboratory test to determine the amount of alkalinity required to raise the pH to the desired level. The results of the lab test must then be converted to the actual quantity required by the digester.

$$\text{Chemical Required, lb} = \frac{\text{Chem. Used in Lab Test, mg} \times \text{Dig. Vol.} \times 3.785}{\text{Sample Vol., L} \times 454\text{ g/lb} \times 1{,}000\text{ mg/g}} \tag{11.3}$$

Example 11.3

Problem:
Around 240 mg of lime will increase the pH of a 1 L sample of the aerobic digester contents to pH 7.1. The digester volume is 240,000 gal. How many pounds of lime will be required to increase the digester pH to 7.3?

Solution:

$$\text{Chemical Required, lb} = \frac{240\text{ mg} \times 240{,}000\text{-gal} \times 3.785\text{ L/gal}}{1\text{L} \times 454\text{ g/lb} \times 1{,}000\text{ mg/g}} = 480\text{ lb}$$

11.3 ANAEROBIC DIGESTION PROCESS CONTROL CALCULATIONS

The purpose of *anaerobic digestion* is the same as aerobic digestion: to stabilize organic matter, to reduce volume, and to eliminate pathogenic organisms. Equipment used in anaerobic digestion includes an anaerobic digester of either the floating- or fixed-cover type (Figure 11.1). These include biosolids pumps for biosolids addition and withdrawal, as well as heating equipment, such as heat exchangers, heaters, and pumps, and mixing equipment for recirculation. Typical ancillaries include gas storage, cleaning equipment, and safety equipment, such as vacuum relief and pressure relief devices, flame traps, and explosion-proof electrical equipment. In the anaerobic process, biosolids enter the sealed digester where organic matter decomposes anaerobically. Anaerobic digestion is a two-stage process:

(1) Sugars, starches, and carbohydrates are converted to volatile acids, carbon dioxide, and hydrogen sulfide.
(2) Volatile acids are converted to methane gas.

Key anaerobic digestion process control calculations are covered in the sections that follow.

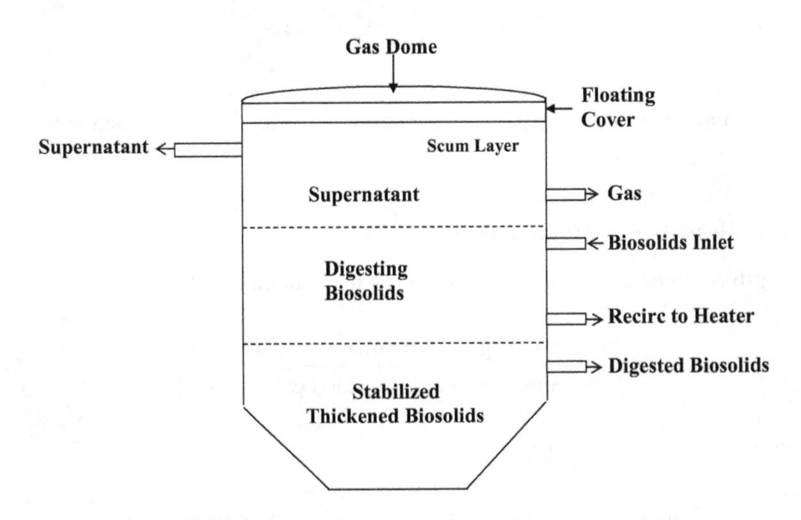

FIGURE 11.1 Floating-roof digester.

11.3.1 Required Seed Volume in Gallons

$$\text{Seed Volume (Gallons)} = \text{Digester Volume, gal} \times \%\ \text{Seed} \tag{11.4}$$

Example 11.4

Problem:
The new digester requires as 25% seed to achieve normal operation within the allotted time. If the digester volume is 280,000 gal, how many gallons of seed material will be required?

Solution:

$$\text{Seed Volume (gal)} = 280{,}000 \times 0.25 = 70{,}000 \text{ gal}$$

11.3.2 Volatile-Acids-to-Alkalinity Ratio

The volatile-acids-to-alkalinity ratio can be used to control the anaerobic digester.

$$\text{Ratio} = \frac{\text{Volatile Acids Concentration}}{\text{Alkalinity Concentration}} \tag{11.5}$$

Example 11.5

Problem:
The digester contains 240 mg/L volatile acids and 1,840 mg/alkalinity. What is the volatile acids/alkalinity ratio?

Solution:

$$\text{Ratio} = \frac{240 \text{ mg/L}}{1{,}840 \text{ mg/L}} = 0.13$$

✓ **Key Point:** Increases in the ratio normally indicate a potential change in the operating condition of the digester.

11.3.3 Biosolids Retention Time

The length of time the biosolids remain in the digester can be expressed as:

$$\text{BRT} = \frac{\text{Digester Volume in Gallons}}{\text{Biosolids Volume added per day, gpd}} \tag{11.6}$$

Example 11.6

Problem:
Biosolids are added to a 520,000 gal digester at the rate of 12,600 gal per day. What is the biosolids retention time?

Solution:

$$\text{BRT} = \frac{520{,}000 \text{ gal}}{12{,}600 \text{ gpd}} = 41.3 \text{ days}$$

11.3.4 Estimated Gas Production in Cubic Feet/Day

The rate of gas production is normally expressed as the volume of gas (in cubic feet) produced per pound of volatile matter destroyed. The total cubic feet of gas a digester will produce per day can be calculated by:

$$\text{Gas Production, ft}^3\text{/day} = \text{Vol. Matter In, lb/day} \times \%\ \text{Vol. Mat. Reduction} \times \text{Prod. Rate ft}^3\text{/lb} \tag{11.7}$$

✓ **Key Point:** Multiplying the volatile matter added to the digester per day by the percent volatile matter reduction (in decimal percent) gives the amount of volatile matter being destroyed by the digestion process per day.

Example 11.7

Problem:
The digester reduces 11,500 lb of volatile matter per day. Currently, the volatile matter reduction achieved by the digester is 55%. The rate of gas production is 11.2 cu ft of gas per pound of volatile matter destroyed.

Solution:

$$\text{Gas Prod.} = 11500 \text{ lb/day} \times 0.55 \times 11.2 \text{ ft}^3\text{/lb} = 70{,}840 \text{ ft}^3\text{/day}$$

11.3.5 Volatile Matter Reduction in Percent

Because of the changes occurring during biosolids digestion, the calculation used to determine percent volatile matter reduction is more complicated.

$$\%\ \text{Red.} = \frac{(\%\ \text{Vol. Matter}_{in} - \%\ \text{Vol. Matter}_{out}) \times 100}{[\%\ \text{Vol. Matter}_{in} - (\%\ \text{Vol. Matter}_{in} \times \%\ \text{Vol. Matter}_{out})]} \tag{11.8}$$

Example 11.8

Problem:
Using the digester data provided here, determine the percent volatile matter reduction for the digester: raw biosolids volatile matter, 71%; digested biosolids volatile matter, 54%.

Solution:

$$\%\ \text{Volatile Matter Reduction} = \frac{0.71 - 0.54}{[0.71 - (0.71 \times 0.54)]} = 52\%$$

11.3.6 Percent Moisture Reduction in Digested Biosolids

$$\% \text{ Moisture Reduction} = \frac{(\% \text{ Moisture}_{in} - \% \text{ Moisture}_{out}) \times 100}{[\% \text{ Moisture}_{in} - (\% \text{ Moisture}_{in} \times \% \text{ Moisture}_{out})]} \quad (11.9)$$

✓ **Key Point:** % moisture = 100%—percent solids

Example 11.9

Problem:

Using the digester data provided here, determine the percent moisture reduction and percent volatile matter reduction for the digester.

Raw biosolids	Percent solids	9%
	Percent moisture	91% (100% – 9%)
Digested biosolids	Percent solids	15%
	Percent moisture	85% (100% – 15%)

Solution:

$$\% \text{ Moisture Reduction} = \frac{(0.91 - 0.85) \times 100}{[0.91 - (0.91 \times 0.85)]} = 44\%$$

11.4 BIOSOLIDS DIGESTION PRACTICE PROBLEMS

Problem 11.1

If 8,000 lb/day of solids with a volatile solids content of 70% are sent to the digester, how many pounds per day volatile solids (VS) are sent to the digester?

Solution:

$$\begin{aligned} \text{Volatile Solids (lb/d)} &= (\text{Biosolids, lb/d})(\% \text{ Solids})(\% \text{ VS}) \\ &= (8{,}000 \text{ lb/d})(0.70) \\ &= 5600 \text{ lb/d} \end{aligned}$$

Problem 11.2

A total of 3,700 gpd of biosolids is pumped to the digester. If the biosolids have 5.8% solids content, with 70% volatile solids, how many pounds per day volatile solids are pumped to the digester?

Solution:

$$\begin{aligned} \text{Volatile Solids (lb/d)} &= (3700 \text{ gpd})(8.34)(0.058)(0.70) \\ &= 1553 \text{ lb/d} \end{aligned}$$

Problem 11.3

What is the digester loading if a digester 40 ft in diameter with a liquid level of 20 ft receives 80,000 lb/day of biosolids with 5.5% solids and 68% volatile solids?

Solution:

$$\text{Digester Loading (lb/d/ft}^3) = \frac{\text{(Biosolids, lb/d)(\% Solids)(\% VS)}}{(0.785)(\text{D, ft})^2\,(\text{depth, ft})}$$

$$= \frac{(80{,}000\text{ lb/d})(0.055)(0.68)}{(0.785)(40\text{ ft})^2\ (20\text{ ft})} = 0.12\text{ lb/d/ft}^3$$

Problem 11.4

The volatile acids (VA) concentration of the biosolids in an anaerobic digester is 170 mg/L. If the measured alkalinity is 2,000 mg/L, what is the VA/alkalinity ratio?

Solution:

$$\text{VA/Alk} = \frac{\text{VA, mg/L}}{\text{Alk, mg/L}} = \frac{170\text{ mg/L}}{2000\text{ mg/L}} = 0.085$$

Problem 11.5

A 50 ft aerobic digester has a side water depth of 12 ft. The biosolids flow to the digester is 9,000 gpd. Calculate the digestion time in days.

Solution:

$$\text{Digestion Time (days)} = \frac{(0.785)(50\text{ ft})^2\,(12\text{ ft})(7.48)}{9000\text{ gpd}}$$

$$= 19.6\text{ days}$$

12 Biosolids Dewatering and Disposal

12.1 BIOSOLIDS DEWATERING

The process of removing enough water from a liquid biosolids to change its consistency to that of damp solid is called *biosolids dewatering*. Although the process is also called *biosolids drying*, the "dry" or dewatered biosolids may still contain a significant amount of water, often as much as 70%. But at moisture contents of 70% or less, the biosolids no longer behave as a liquid and can be handled manually or mechanically. Several methods are available to dewater biosolids. The particular types of dewatering techniques/devices used best describe the actual processes used to remove water from biosolids and change their form from a liquid to damp solid. The commonly used techniques/devices include the following:

- Filter presses
- Vacuum filtration
- Sand drying beds

✓ **Key Point:** Centrifugation is also used in the dewatering process. However, in this text we concentrate on those unit processes listed previously that are traditionally used for biosolids dewatering.

Note that an ideal dewatering operation would capture all the biosolids at minimum cost, and the resultant dry biosolids solids or cake would be capable of being handled without causing unnecessary problems. Process reliability, ease of operation, and compatibility with the plant environment would also be optimized.

12.2 PRESSURE FILTRATION CALCULATIONS

In *pressure filtration*, the liquid is forced through the filter media by a positive pressure. Several types of presses are available, but the most commonly used types are plate and frame presses and belt presses.

12.2.1 Plate and Frame Press

The *plate and frame press* consists of vertical plates that are held in a frame and that are pressed together between a fixed and moving end. A cloth filter medium is mounted on the face of each individual plate. The press is closed, and biosolids are pumped into the press at pressures up to 225 psi and pass through feed holes in the trays along the length of the press. Filter presses usually require a precoat

DOI: 10.1201/9781003354314-12

material, such as incinerator ash or diatomaceous earth, to aid in solids retention on the cloth and to allow easier release of the cake. Performance factors for plate and frame presses include feed biosolids characteristics, type and amount of chemical conditioning, operating pressures, and the type and amount of precoat. Filter press calculations (and other dewatering calculations) typically used in wastewater solids handling operations include solids loading rate, net filter yield, hydraulic loading rate, biosolids feed rate, solids loading rate, flocculant feed rate, flocculant dosage, total suspended solids, and percent solids recovery.

12.2.1.1 Solids Loading Rate

The solids loading rate is a measure of the pounds per hour solids applied per square foot of plate area, as shown in equation 12.1.

$$\text{Sol. Loading Rate, lb/hr/sq ft} = \frac{\text{(Biosolids, gph) (8.34, lb/gal)(\% Sol./100)}}{\text{Plate Area, sq ft}} \quad (12.1)$$

Example 12.1

Problem:

A filter press used to dewater digested primary biosolids receives a flow of 710 gal during a 2 hr period. The biosolids has a solids content of 3.3%. If the plate surface area is 120 sq ft, what is the solids loading rate in pounds per hour per square foot?

Solution:

The flow rate is given as gallons per 2 hr. First, express this flow rate as gallons per hour: 710 gal/2 hr = 355 gal/hr.

$$\text{Solids Loading Rate, lb/hr/sq ft} = \frac{\text{(Biosolids, gph) (8.34 lb/gal)} \ \frac{\text{(\% Solids)}}{100}}{\text{Plate Area, sq ft}}$$

$$= \frac{\text{(355 gph) (8.34 lb/gal)} \ \frac{(3.3)}{100}}{120\ \text{sq ft}}$$

$$= 0.81\ \text{lb/hr/sq ft}$$

✓ **Key Point:** The solids loading rate measures the pounds per hour of solids applied to each square foot of plate surface area. However, this does not reflect the time when biosolids feed to the press is stopped.

12.2.1.2 Net Filter Yield

Operated in the batch mode, biosolids are fed to the plate and frame filter press until the space between the plates is completely filled with solids. The biosolids flow to the press is then stopped and the plates are separated, allowing the biosolids cake to fall into a hopper or conveyor below. The *net filter yield*, measured in pounds per hour per square foot, reflects the run time as well as the downtime of the plate and frame filter press. To calculate the net filter yield, simply multiply the solids loading rate (in pounds per hour per square foot) by the ratio of filter run time to total cycle time as follows:

$$\text{Net Filter Yield.} = \frac{(\text{Biosolids, gph})\ (8.34\ \text{lb/gal})\ (\%\ \text{Sol/100})}{\text{Plate Area, sq ft}} \times \frac{\text{Filter Run Time}}{\text{Total Cycle Time}} \quad (12.2)$$

Example 12.2

Problem:
A plate and frame filter press receives a flow of 660 gal of biosolids during a 2 hr period. The solids concentration of the biosolids is 3.3%. The surface area of the plate is 110 sq ft. If the downtime for biosolids cake discharge is 20 min, what is the net filter yield in pounds per hour per square foot?

Solution:
First, calculate the solids loading rate, then multiply that number by the corrected time factor:

$$\text{Solids Loading Rate} = \frac{(\text{Biosolids, gph})\ (8.34\ \text{lb/gal})\ (\%\ \text{Sol./100})}{\text{Plate Area, sq ft}}$$

$$= \frac{(330\ \text{gph})\ (8.34\ \text{lb/gal})\ (3.3/100)}{100\ \text{sq ft}}$$

$$= 0.91\ \text{lb/hr/sq ft}$$

Next, calculate net filter yield using the corrected time factor:

$$\text{Net Filter Yield, lb/hr/sq ft} = \frac{(0.91\ \text{lb/hr/sq ft})\ (2\ \text{hr})}{2.33\ \text{hr}}$$

$$= 0.78\ \text{lb/hr/sq ft}$$

12.2.2 Belt Filter Press

The *belt filter press* consists of two porous belts. The biosolids are sandwiched between the two porous belts (see Figure 12.1). The belts are pulled tight together as they are passed around a series of rollers to squeeze water out of the biosolids. Polymer is added to the biosolids just before it gets to the unit. The biosolids is then distributed across one of the belts to allow for some of the water to drain by gravity. The belts are then put together with the biosolids in between.

12.2.3 Hydraulic Loading Rate

Hydraulic loading for belt filters is a measure of gallons per minute flow per foot or belt width. Figure 12.2 and the associated equation are shown here:

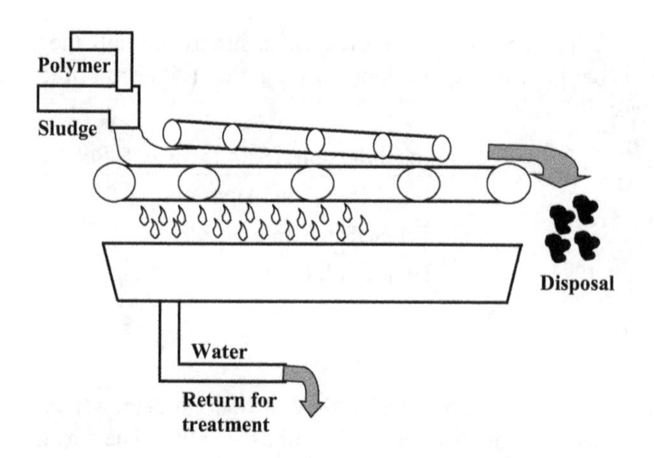

FIGURE 12.1 Belt filter press.

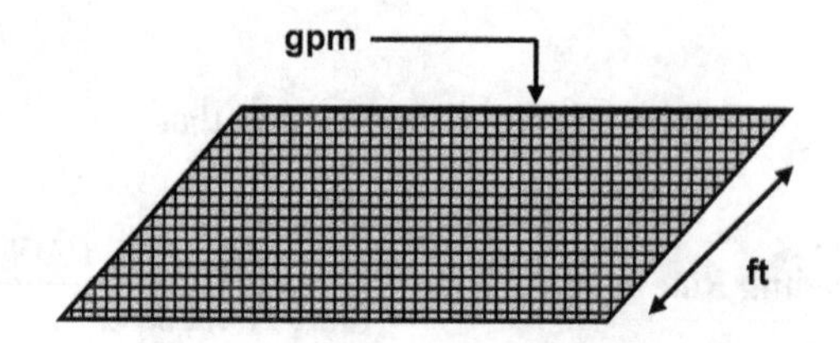

FIGURE 12.2 Belt filter press hydraulic loading rate.

$$\text{Hydraulic Loading Rate, gpm/ft} = \frac{\text{Flow, gpm}}{\text{Belt Width, ft}} \tag{12.3}$$

Example 12.3

Problem:
A 6 ft wide belt press receives a flow of 110 gpm of primary biosolids. What is the hydraulic loading rate in gallons per minute per foot?

Solution:

$$\begin{aligned}\text{Hydraulic Loading Rate, gpm/ft} &= \frac{\text{Flow, gpm}}{\text{Belt Width, ft}}\\ &= \frac{110\text{ gpm}}{6\text{ ft}}\\ &= 18.3\text{ gpm/ft}\end{aligned}$$

Example 12.4

Problem:
A belt filter press 5 ft wide receives a primary biosolids flow of 150 gpm. What is the hydraulic loading rate in gallons per minute per square foot?

Solution:

$$\text{Hydraulic Loading Rate, gpm/ft} = \frac{\text{Flow, gpm}}{\text{Belt Width, ft}}$$

$$= \frac{150 \text{ gpm}}{5 \text{ ft}}$$

$$= 30 \text{ gpm/ft}$$

12.2.4 Biosolids Feed Rate

The biosolids feed rate to the belt filter press depends on several factors, including the biosolids, in pounds per day, that must be dewatered; the maximum solids feed rate, in pounds per hour, that will produce an acceptable cake dryness; and the number of hours per day the belt press is in operation. The equation used in calculating biosolids feed rate is:

$$\text{Biosolids Feed Rate, lb/hr} = \frac{\text{Biosolids to be dewatered, lb/day}}{\text{Operating Time, hr/day}} \quad (12.4)$$

Example 12.5

Problem:
The amount of biosolids to be dewatered by the belt filter press is 20,600 lb/day. If the belt filter press is to be operated 10 hr each day, what should the biosolids feed rate in pounds per hour be to the press?

Solution:

$$\text{Biosolids Feed Rate, lb/hr} = \frac{\text{Biosolids to be dewatered, lb/day}}{\text{Operating Time, hr/day}}$$

$$= \frac{20{,}600 \text{ lb/day}}{10 \text{ hr/day}}$$

$$= 2060 \text{ lb/hr}$$

12.2.5 Solids Loading Rate

The solids loading rate may be expressed as pounds per hour or as tons per hour. In either case, the calculation is based on biosolids flow (or feed) to the belt press and percent of milligrams per liter concentration of total suspended solids (TSS) in the biosolids. The equation used in calculating solids loading rate is:

$$\text{Sol. Load. Rate, lb/hr} = (\text{Feed, gpm})\ (60 \text{ min/hr})\ (8.34 \text{ lb/gal})(\% \text{ TSS}/100) \quad (12.5)$$

Example 12.6

Problem:
The biosolids feed to a belt filter press is 120 gpm. If the total suspended solids concentration of the feed is 4%, what is the solids loading rate, in pounds per hour?

Solution:

$$\text{Sol. Load. Rate, lb/hr} = (\text{Feed, gpm})\,(60\text{ min/hr})\,(8.34\text{ lb/gal})(\%\text{ TSS})/100$$
$$= (120\text{ gpm})\,(60\text{ min/hr})\,(8.34\text{ lb/gal})\,(4/100)$$
$$= 2402\text{ lb/hr}$$

12.2.6 Flocculant Feed Rate

The flocculant feed rate may be calculated like all other milligrams per liter to pounds per day calculations and then converted to pounds per hour feed rate, as follows:

$$\text{Flocculant Feed, lb/day} = \frac{(\text{Floc., mg/L})\,(\text{Feed Rate, MGD})\,(8.34\text{ lb/gal})}{24\text{ hr/day}} \tag{12.6}$$

Example 12.7

Problem:
The flocculant concentration for a belt filter press is 1% (10,000 mg/L). If the flocculant feed rate is 3 gpm, what is the flocculant feed rate in pounds per hour?

Solution:
First, calculate pounds per day flocculant using the milligrams per liter to pounds per day calculation. Note that the gallons per minute feed flow must be expressed as million gallons per day feed flow:

$$\frac{(3\text{ gpm})\,(1440\text{ min/day})}{1{,}000{,}000} = 0.00432\text{ MGD}$$

$$\text{Flocculant Feed, lb/day} = (\text{mg/L Floc.})\,(\text{Feed Rate, MGD})\,(8.34\text{ lb/gal})$$
$$= (10{,}000\text{ mg/L})\,(0.00432\text{ MGD})\,(8.34\text{ lb/gal})$$
$$= 360\text{ lb/day}$$

Then, convert pounds per day flocculant to pounds per hour:

$$= \frac{360\text{ lb/day}}{24\text{ hr/day}} = 15\text{ lb/hr}$$

12.2.7 Flocculant Dosage

Once the solids loading rate (tons per hour) and flocculant feed rate (pounds per hour) have been calculated, the flocculant dose in pounds per ton can be determined. The equation used to determine flocculant dosage is:

$$\text{Flocculant Dosage, lb/ton} = \frac{\text{Flocculant, lb/hr}}{\text{Solids Treated, ton/hr}} \tag{12.7}$$

Example 12.8

Problem:
A belt filter has solids loading rate of 3,100 lb/hr and a flocculant feed rate of 12 lb/hr. Calculate the flocculant dose in pounds per ton of solids treated.

Solution:
First, convert pounds per hour solids loading to tons per hour solids loading:

$$\frac{3100 \text{ lb/hr}}{2000 \text{ lb/ton}} = 1.55 \text{ ton/hr}$$

Now, calculate pound flocculant per ton of solids treated:

$$\text{Flocculant Dosage, lb/ton} = \frac{\text{Flocculant, lb/hr}}{\text{Solids Treated, ton/hr}}$$

$$= \frac{12 \text{ lb/hr}}{1.55 \text{ ton/hr}}$$

$$= 7.7 \text{ lb/ton}$$

12.2.8 Total Suspended Solids

The feed biosolids solids are comprised of two types of solids: suspended solids and dissolved solids (see Figure 12.3).

- *Suspended solids* will not pass through a glass fiber filter pad. They can be further classified as total suspended solids (TSS), volatile suspended solids, and/or fixed suspended solids. They can also be separated into three components based on settling characteristics: settleable solids, floatable solids, and colloidal solids. Total suspended solids in wastewater is normally in the range of 100–350 mg/L.
- *Dissolved solids* will pass through a glass fiber filter pad. They can also be classified as total dissolved solids (TDS), volatile dissolved solids, and fixed dissolved solids. Total dissolved solids is normally in the range of 250–850 mg/L.

Two lab tests can be used to estimate the total suspended solids concentration of the feed biosolids concentration of the feed biosolids to the filter press: total residue test (measures both suspended and dissolved solids concentrations) and total filterable residue test (measures only the dissolved solids concentration). By subtracting

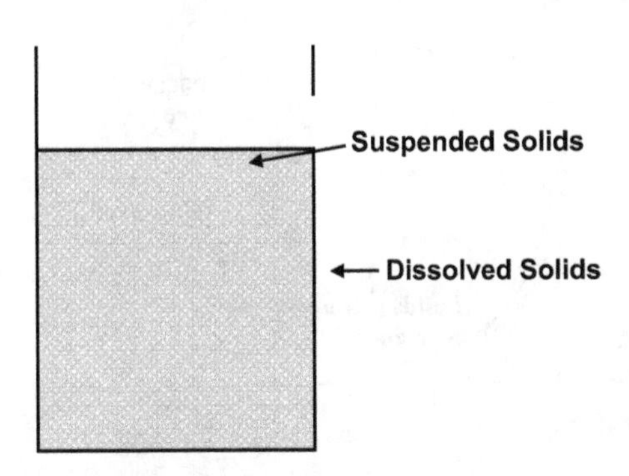

FIGURE 12.3 Feed biosolids solids.

the total filterable residue from the total residue, the result is the total non-filterable residue (total suspended solids), as shown in equation (12.8).

$$\text{Total Res., mg/L} - \text{Total Filterable Residue, mg/L} = \text{Total Non-Filterable Residue, mg/L} \quad (12.8)$$

Example 12.9

Problem:
Lab tests indicate that the total residue portion of a feed biosolids sample is 22,000 mg/L. The total filterable residue is 720 mg/L. On this basis, what is the estimated total suspended solids concentration of the biosolids sample?

Solution:

$$\text{Total Residue, mg/L} - \text{Total Filterable Res., mg/L} = \text{Total Non-Filterable Res., mg/L}$$

$$22{,}000 \text{ mg/L} - 720 \text{ mg/L} = 21{,}280 \text{ mg/L Total SS}$$

12.3 ROTARY VACUUM FILTER DEWATERING CALCULATIONS

The *rotary vacuum filter* (see Figure 12.4) is a device used to separate solid material from liquid. The vacuum filter consists of a large drum with large holes in it covered with a filter cloth. The drum is partially submerged and rotated through a vat of conditioned biosolids. Capable of excellent solids capture and high-quality supernatant/filtrate, solids concentrations of 15–40% can be achieved.

12.3.1 Filter Loading

The filter loading for vacuum filters is a measure of pounds per hour of solids applied per square foot of drum surface area. The equation to be used in this calculation is shown here:

$$\text{Filter Loading, lb/hr/sq ft} = \frac{\text{Solids to Filter, lb/hr}}{\text{Surface Area, sq ft}}$$

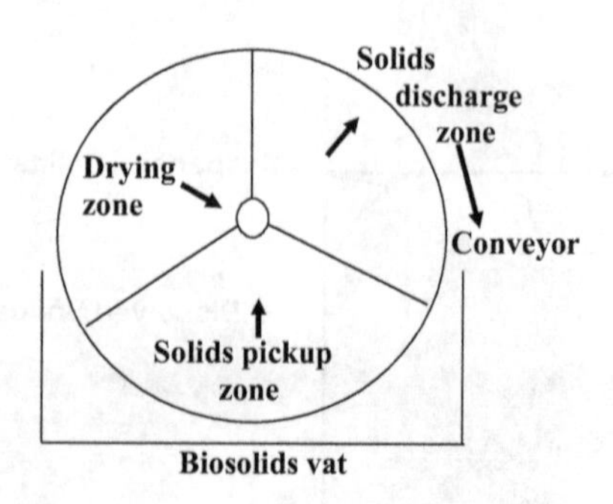

FIGURE 12.4 Vacuum filter.

Example 12.10

Problem:
Digested biosolids is applied to a vacuum filter at a rate of 70 gpm, with a solids concentration of 3%. If the vacuum filter has a surface area of 300 sq ft, what is the filter loading in pounds per hour per square foot?

Solution:

$$\text{Filter Loading, lb/hr/sq ft} = \frac{(\text{Biosolids, gpm})\,(60\text{ min/hr})\,(8.34\text{ lb/gal})\,(\%\text{ Sol.}/100)}{\text{Surface Area, sq ft}}$$

$$= \frac{(70\text{ gpm})\,(60\text{ min/hr})\,(8.34\text{ lb/gal})\,(3/100)}{300\text{ sq ft}}$$

$$= 3.5\text{ lb/hr/sq ft}$$

12.3.2 Filter Yield

One of the most common measures of vacuum filter performance is filter yield. It is the pounds per hour of dry solids in the dewatered biosolids (cake) discharged per square foot of filter area. It can be calculated using equation 12.10.

$$\text{Filter Yield, lb/r/sq ft} = \frac{(\text{Wet Cake Flow, lb/hr})\,\dfrac{(\%\text{ Solids in Cake})}{100}}{\text{Filter Area, sq ft}} \quad (12.10)$$

Example 12.11

Problem:
The wet cake flow from a vacuum filter is 9,000 lb/hr. If the filter area is 300 sq ft and the percent solids in the cake is 25%, what is the filter yield in pounds per hour per square foot?

Solution:

$$\text{Filter Area, sq ft} = \frac{(\text{Wet Cake Flow, lb/hr})\,\dfrac{(\%\text{ Solids in Cake})}{100}}{\text{Filter Area, sq ft}}$$

$$= \frac{(9000\text{ lb/hr})\,\dfrac{(25)}{100}}{300\text{ sq ft}}$$

$$= 7.5\text{ lb/hr sq ft}$$

12.3.3 Vacuum Filter Operating Time

The vacuum filter operating time required to process a given pounds per day solids can be calculated using equation 12.10. However, the vacuum filter operating time, of course, is the unknown factor, designated by *x*.

Example 12.12

Problem:
A total of 4,000 lb/day primary biosolids solids are to be processed by a vacuum filter. The vacuum filter yield is 2.2 lb/hr/sq ft. The solids recovery is 95%. If the area of the filter is 210 sq ft, how many hours per day must the vacuum filter remain in operation to process these solids?

Solution:

$$\text{Filter Yield, lbs/hr/sq ft} = \frac{\dfrac{\text{Solids To Filter, lb/day}}{\text{Filter Operation., lb/day}}}{\text{Filter Area, sq ft}} \times \frac{(\%\text{ Recovery})}{100}$$

$$\text{2.2-lb/hr/sq ft} = \frac{\dfrac{4000\text{ lb/day}}{x\text{ hr/day Oper.}}}{210\text{ sq ft}} \times \frac{95}{100}$$

$$\text{2.2-lb/hr/sq ft} = \frac{(4000\text{ lb/day})}{x\text{ hr/day}} \times \frac{(1)}{210\text{ sq ft}} \frac{(95)}{100}$$

$$x = \frac{(4000)(1)(95)}{(2.2)(210)(100)}$$

$$x = 8.2\text{ hr/day}$$

12.3.4 Percent Solids Recovery

As mentioned, the function of the vacuum filtration process is to separate the solids from the liquids in the biosolids being processed. Therefore, the percent of feed solids "recovered" (sometimes referred to as the percent solids capture) is a measure of the efficiency of the process. Equation 12.11 is used to determine percent solids recovery.

$$\%\text{ Sol Rec.} = \frac{(\text{Wet Cake Flow, lb/hr})\dfrac{(\%\text{ Sol. In Cake})}{100}}{(\text{Biosolids Feed, lbs/hr})\dfrac{(\%\text{ Sol. In Feed})}{100}} \times 100 \qquad (12.11)$$

Example 12.13

Problem:
The biosolids fed to a vacuum are 3,400 lb/day, with a solids content of 5.1%. If the wet cake flow is 600 lb/hr with 25% solids content, what is the percent solids recovery?

Solution:

$$\%\text{ Sol. Rec.} = \frac{(\text{Wet Cake Flow, lb/hr})\dfrac{(\%\text{ Sol. In Cake})}{100}}{(\text{Biosolids Feed, lb/hr})\dfrac{(\%\text{ Sol. In Feed})}{100}} \times 100$$

$$= \frac{(600\text{ lb/hr})\,\frac{(25)}{100}}{(3400\text{ lb/hr})\,\frac{(5.1)}{100}} \times 100$$

$$= \frac{150\text{ lb/hr}}{173\text{ lb/hr}} \times 100$$

$$\%\text{ Sol. Rec.} = 87\ \%$$

12.4 SAND DRYING BED CALCULATIONS

Drying beds are generally used for dewatering well-digested biosolids. Biosolids drying beds consist of a perforated or open joint drainage system in a support media, usually gravel or wire mesh. Drying beds are usually separated into workable sections by wood, concrete, or other materials. Drying beds may be enclosed or opened to the weather. They may rely entirely on natural drainage and evaporation processes or may use a vacuum to assist the operation. *Sand drying beds* are the oldest biosolids dewatering technique and consist of 6–12 in of coarse sand underlain by layers of graded gravel ranging from 1/8 to 1/4 in at the top and 3/4 to 1 1/2 in of the bottom. The total gravel thickness is typically about 1 ft. Graded natural earth (4 to 6 in) usually makes up the bottom, with a web of drain tile placed on 20–30 ft centers. Sidewalls and partitions between bed sections are usually of wooden planks or concrete and extend about 14 in above the sand surface.

12.4.1 SAND DRYING BEDS PROCESS CONTROL CALCULATIONS

Typically, three calculations are used to monitor sand drying bed performance—total biosolids applied, solids loading rate, and biosolids withdrawal to drying beds.

12.4.2 TOTAL BIOSOLIDS APPLIED

The total gallons of biosolids applied to sand drying beds may be calculated using the dimensions of the bed and the depth of biosolids applied, as shown by equation 12.12.

$$\text{Volume, gal} = (\text{length, ft})\ (\text{width, ft})\ (\text{depth, ft})\ (7.48,\text{ gal/cu ft}) \tag{12.12}$$

Example 12.14

Problem:
A drying bed is 220 ft long and 20 ft wide. If biosolids are applied to a depth of 4 in, how many gallons of biosolids are applied to the drying bed?

Solution:

$$\begin{aligned}\text{Volume, gal} &= (\text{l})\ (\text{w})\ (\text{d})\ (7.48\text{ gal/cu ft})\\ &= (220\text{ ft})\ (20\text{ ft})\ (0.33\text{ ft})\ (7.48\text{ gal/cu ft})\\ &= 10{,}861\text{ gal}\end{aligned}$$

12.4.3 Solids Loading Rate

The biosolids loading rate may be expressed as pounds per year per square foot. The loading rate is dependent on biosolids applied per applications, pounds, percent solids concentration, cycle length, and square feet of sand bed area. The equation for biosolids loading rate is given here:

$$\text{Sol. Load. Rate, lbs/yr/sq ft} = \frac{\dfrac{\text{lb Biosolids Applied}}{\text{Days of Application}}(365\text{ days/yr})\dfrac{(\%\text{ Solids})}{100}}{(\text{length, ft})(\text{width, ft})} \qquad (12.13)$$

Example 12.15

Problem:
A biosolids bed is 210 ft long and 25 ft wide. A total of 172,500 lb of biosolids is applied each application of the sand drying bed. The biosolids have a solids content of 5%. If the drying-and-removal cycle requires 21 days, what is the solids loading rate in pounds per year per square foot?

Solution:

$$\text{Sol. Load. Rate, lbs/yr/sq ft} = \frac{\dfrac{\text{lb Biosolids Applied}}{\text{Days of Application}}\times(365\text{ days/yr})\times\dfrac{(\%\text{ Solids})}{100}}{\text{Bed Area, sq ft}}$$

$$= \frac{\dfrac{(172{,}500\text{ lb})}{21\text{ days}}\times(365\text{ day/yr})\times\dfrac{(5)}{100}}{(210\text{ ft})(25\text{ ft})}$$

$$= 28.6\text{ lb/yr/sq ft}$$

12.4.4 Biosolids Withdrawal to Drying Beds

Pumping digested biosolids to drying beds is one method among many for dewatering biosolids, thus making the dried biosolids useful as a soil conditioner. Depending upon the climate of a region, the drying bed depth may range from 8 to 18 in. Therefore, the area covered by these drying beds may be substantial. For this reason, the use of drying beds is more common for smaller plants than for larger plants. Use the following to calculate biosolids withdrawal to drying beds:

$$\text{Biosolids Withdrawn, cu ft} = (0.785)(D^2)(\text{Drawdown, ft}) \qquad (12.14)$$

Example 12.16

Problem:
Biosolids is withdrawn from a digester that has a diameter of 40 ft. If the biosolids is drawn down 2 ft, how many cubic feet will be sent to the drying beds?

Solution:

$$\text{Biosolids Withdrawal, cu ft} = (0.785)\,(D^2)\,(\text{ft drop})$$
$$= (0.785)\,(40\text{ ft})\,(40\text{ ft})\,(2\text{ ft})$$
$$= 2512\text{ cu ft withdrawn}$$

12.5 BIOSOLIDS DISPOSAL

In the disposal of biosolids, land application, in one form or another, has become not only a necessity (because of the banning of ocean dumping in the United States in 1992 and the shortage of landfill space since then) but also quite popular as a beneficial reuse practice. That is, beneficial reuse means that the biosolids are disposed of in an environmentally sound manner by recycling nutrients and soil conditions. Biosolids are being applied throughout the United States to agricultural and forest lands. For use in land applications, the biosolids must meet certain conditions. Biosolids must comply with state and federal biosolids management/disposal regulations and must also be free of materials dangerous to human health (i.e., toxicity, pathogenic organisms, etc.) and/or dangerous to the environment (i.e., toxicity, pesticides, heavy metals, etc.). Biosolids are land applied by direct injection, by application and incorporation (plowing in), or by composting.

12.5.1 Land Application Calculations

Land application of biosolids requires precise control to avoid problems. Use of process control calculations is part of the overall control process. Calculations include determining disposal cost, plant available nitrogen (PAN), application rate (dry tons and wet tons per acre), metals loading rates, maximum allowable applications based upon metals loading, and site life based on metals loading.

12.5.2 Disposal Cost

The cost of disposal of biosolids can be determined by:

$$\text{Cost} = \text{Wet Tons Biosolids Produced/Year} \times \%\ \text{Solids} \times \text{Cost/dry ton} \qquad (12.15)$$

Example 12.17

Problem:
The treatment system produces 1,925 wet tons of biosolids for disposal each year. The biosolids are 18% solids. A contractor disposes of the biosolids for $28 per dry ton. What is the annual cost for biosolids disposal?

Solution:

$$\text{Cost} = 1{,}925\text{ wet tons/year} \times 0.18 \times \$2800\text{/dry ton} = \$9{,}702$$

12.5.2.1 Plant Available Nitrogen (PAN)

One factor considered when land applying biosolids is the amount of nitrogen in the biosolids available to the plants grown on the site. This includes ammonia nitrogen and organic nitrogen. The organic nitrogen must be mineralized for plant consumption. Only a portion of the organic nitrogen is mineralized per year. The mineralization factor (f^1) is assumed to be 0.20. The amount of ammonia nitrogen available is directly related to the time elapsed between applying the biosolids and incorporating (plowing) the biosolids into the soil. We provide volatilization rates based upon the example that follows.

$$\text{PAN, lb/dry ton} = [(\text{Organic Nitrogen, mg/kg} \times f^1) + (\text{Ammonium Nitrate. mg/kg} \times \text{V1})] \times 0.002 \text{ lb/dry ton} \quad (12.16)$$

Where:

F^1 = mineral rate for organic nitrogen (assume 0.20)
V_1 = volatilization rate ammonia nitrogen
V_1 = 1.00 if biosolids are injected
V_1 = 0.85 if biosolids are plowed in within 24 hr
V_1 = 0.70 if biosolids are plowed in within 7 days

Example 12.18

Problem:
The biosolids contain 21,000 mg/kg of organic nitrogen and 10,500 mg/kg of ammonia nitrogen. The biosolids are incorporated into the soil within 24 hr after application. What is the plant available nitrogen (PAN) per dry ton of solids?

Solution:

$$\text{PAN, lb/dry ton} = [(21{,}000 \text{ mg/kg} \times 0.20) + (10{,}500 \times 0.85)] \times 0.002$$
$$= 26.3 \text{ lb PAN/dry ton}$$

12.5.3 Application Rate Based on Crop Nitrogen Requirement

In most cases, the application rate of domestic biosolids to crop lands will be controlled by the amount of nitrogen the crop requires. The biosolids application rate based upon the nitrogen requirement is determined by the following:

(1) Use an agriculture handbook to determine the nitrogen requirement of the crop to be grown.
(2) Determine the amount of biosolids in dry tons required to provide this much nitrogen.

$$\text{Dry tons/acre} = \frac{\text{Plant Nitrogen Requirement, lb/acre}}{\text{Plant Available Nitrogen, lb/dry ton}} \quad (12.17)$$

Example 12.19

Problem:
The crop to be planted on the land application site requires 150 lb of nitrogen per acre. What is the required biosolids application rate if the PAN of the biosolids is 30 lb/dry ton?

Solution:

$$\text{Dry tons/acre} = \frac{\text{150 lb nitrogen nitrogen/acre}}{\text{30 lb/dry ton}} = \text{5 dry tons/acre}$$

12.5.4 Metals Loading

When biosolids are land applied, metals concentrations are closely monitored and their loading on land application sites is calculated.

$$\text{Loading, lb/acre} = \text{Metal Concentration., mg/kg} \times 0.002 \text{ lb/dry ton} \times \text{Appl. Rate, dry ton/acre} \quad (12.18)$$

Example 12.20

Problem:
The biosolids contain 14 mg/kg of lead. Biosolids are currently being applied to the site at a rate of 11 dry ton per acre. What is the metals loading rate for lead in pounds per acre?

Solution:

$$\text{Loading Rate, lb/acre} = 14 \text{ mg/kg} \times 0.002 \text{ lb/dry ton} \times 11 \text{ dry ton} = 0.31 \text{ lb/acre}$$

12.5.5 Maximum Allowable Applications Based upon Metals Loading

If metals are present, they may limit the total number of applications a site can receive. Metals loadings are normally expressed in terms of the maximum total amount of metal that can be applied to a site during its use.

$$\text{Applications} = \frac{\text{Max. Allowable Cumulative Load for the Metal, lb/Ac}}{\text{Metal Loading, lb/acre/application}} \quad (12.19)$$

Example 12.21

Problem:
The maximum allowable cumulative lead loading is 48 lb/acre. Based upon the current loading of 0.35 lb/acre, how many applications of biosolids can be made to this site?

Solution:

$$\text{Applications} = \frac{\text{48.0 lb/acre}}{\text{0.35 lb/acre}} = 137 \text{ applications}$$

12.5.6 Site Life Based on Metals Loading

The maximum number of applications based upon metals loading and the number of applications per year can be used to determine the maximum site life.

$$\text{Site Life, years} = \frac{\text{Maximum Allowable Applications}}{\text{Number of Applications Planned/Year}} \tag{12.20}$$

Example 12.22

Problem:
Biosolids are currently applied to a site twice annually. Based upon the lead content of the biosolids, the maximum number of applications is determined to be 135 applications. Based upon the lead loading and the applications rate, how many years can this site be used?

Solution:

$$\text{Site Life} = \frac{135 \text{ applications}}{2 \text{ applications/year}} = 68 \text{ year} \tag{12.21}$$

✓ **Key Point:** When more than one metal is present, the calculations must be performed for each metal. The site life would then be the lowest value generated by these calculations.

12.6 BIOSOLIDS TO COMPOST

The purpose of composting biosolids is to stabilize the organic matter, reduce volume, eliminate pathogenic organisms, and produce a product that can be used as a soil amendment or conditioner. Composting is a biological process. In a composting operation, dewatered solids are usually mixed with a bulking agent (i.e., hardwood chips) and stored until biological stabilization occurs (see Figure 12.5). The composting mixture is ventilated during storage to provide sufficient oxygen for oxidation and to prevent odors. After the solids are stabilized, they are separated from the bulking agent. The composted solids are then stored for curing and are applied to farmlands or other beneficial uses. Expected performance of the composting operation for both percent volatile matter reduction and percent moisture reduction ranges from 40 to 60%.

Performance factors related to biosolids composting include moisture content, temperature, pH, nutrient availability, and aeration. The biosolids must contain sufficient moisture to support the biological activity. If the moisture level is too low (40% less), biological activity will be reduced or stopped. At the same time, if the moisture level exceeds approximately 60%, it will prevent sufficient airflow through the mixture.

The composting process operates best when the temperature is maintained within an operating range of 130–140°F—biological activities provide enough heat to increase the temperature well above this range. Forced air ventilation or mixing is used to remove heat and maintain the desired operating temperature range. The temperature of the composting solids when maintained at the required levels will be sufficient to remove pathogenic organisms.

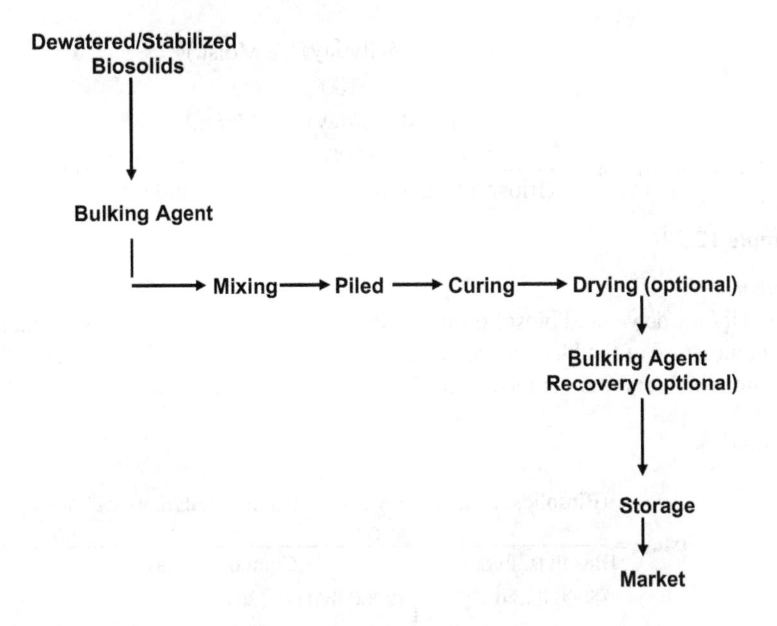

FIGURE 12.5 Flow diagram for composting biosolids.

The influent pH can affect the performance of the process if extreme (less than 6.0 or greater than 11.0). The pH during composting may have some impact on the biological activity but does not appear to be a major factor. Composted biosolids generally have a pH in the range of 6.8 to 7.5.

The critical nutrient in the composting process is nitrogen. The process works best when the ratio of nitrogen to carbon is in the range of 26 to 30 carbon to 1 nitrogen. Above this ratio, composting is slowed. Below this ratio, the nitrogen content of the final product may be less attractive as compost.

Aeration is essential to provide oxygen to the process and to control the temperature. In force air processes, some means of odor control should be included in the design of the aeration system.

12.6.1 Composting Calculations

Pertinent composting process control calculations include determination of percent of moisture of compost mixture and compost site capacity.

12.6.1.1 Blending Dewatered Biosolids with Composted Biosolids

When blending composted material with dewatered biosolids, it is similar to blending two different percent solids biosolids. The percent solids (or percent moisture) content of the mixture will always fall somewhere between the percent solids (or percent moisture) concentrations of the two materials being mixed. Equation 12.22 is used to determine percent moisture of mixture.

$$\% \text{ Moist. of Mixture} = \frac{\frac{(\text{Biosolids, lb/day})(\% \text{ Moist.})}{100} + \frac{(\text{Compost, lb/day})(\% \text{ Moist.})}{100}}{(\text{Biosolids, lb/day}) + (\text{Compost, lb/day})} \times 100 \quad (12.22)$$

Example 12.23

Problem:
If 5,000 lb/day dewatered biosolids are mixed with 2,000 lb/day compost, what is the percent moisture of the blend? The dewatered biosolids have a solids content of 25% (75% moisture), and the compost has a 30% moisture content.

Solution:

$$\% \text{ Moist of Mixture} = \frac{(\text{Biosolids, lb/day})\frac{(\% \text{ Moist.})}{100} + (\text{Compost, lb/day})\frac{(\% \text{ Moist.})}{100}}{\text{Biosolids, lb/day} + \text{Compost, lb/day}} \times 100$$

$$= \frac{(5000 \text{ lb/day})\frac{(75)}{100} + (2000 \text{ lb/day})\frac{(30)}{100}}{5000 \text{ /lb/day} + 2000 \text{ lb/day}}$$

$$= \frac{3750 \text{ lb/day} + 600 \text{ lb/day}}{7000 \text{ lb/day}}$$

$$= 62\%$$

12.6.1.2 Compost Site Capacity Calculation

An important consideration in compost operation is the solids processing capability (fill time), in pounds per day or pounds per week. Equation 12.23 is used to calculate site capacity.

$$\text{Fill Time, days} = \frac{\text{total Available Capacity, cu yds}}{\frac{\text{wet Compost, lb/day}}{\text{compost Bulk Density, lb/cu yd}}} \quad (12.23)$$

Example 12.24

Problem:
A composting facility has an available capacity of 7,600 cubic yards. If the composting cycle is 21 days, how many pounds per day wet compost can be processed by this facility? Assume a compost bulk density of 900 lb/cu yd.

Solution:

$$\text{Fill Time, days} = \frac{\text{Total Available Capacity, cu yds}}{\frac{\text{Wet Compost, lb/day}}{\text{Compost Bulk Density, lb/cu yd}}}$$

$$21 \text{ days} = \frac{7600 \text{ cu yds}}{\frac{x \text{ lb/day}}{900 \text{ lb/cu yd}}}$$

$$21 \text{ days} = \frac{(7600 \text{ cu yds})(900 \text{ lb/cu yd})}{x \text{ lb/day}}$$

$$x \text{ lb/day} = \frac{(7600 \text{ cu yds})(900 \text{ lb/cu yd})}{21 \text{ days}}$$

$$x = 325{,}714 \text{ lb/day}$$

12.7 BIOSOLIDS DEWATERING AND DISPOSAL PRACTICE PROBLEMS

Problem12.1

A filter press used to dewater digested primary biosolids receives a flow of 700 gal during a 2 hr period. The biosolids have a solids content of 3.1%. If the plate surface area is 125 sq ft, what is the solids loading rate in pounds per hour per square foot?

Solution:
The flow rate is given as gallons per 2 hr. First, express this flow rate as gallons per hour: 700 gal/2 hrs = 350 gal/hr.

$$\text{Solids Loading Rate, lb/hr/sq ft} = \frac{(\text{Biosolids, gph})(8.34 \text{ lb/gal})\left(\frac{\% \text{ Solids}}{100}\right)}{\text{Plate Area, sq ft}}$$

$$= \frac{(350 \text{ gph})(8.34 \text{ lb/gal})\left(\frac{3.1}{100}\right)}{125 \text{ sq ft}}$$

$$= 0.72 \text{ lb/hr/sq ft}$$

Problem 12.2

A plate-and-frame filter press receives a flow of 600 gal of biosolids during a 2 hr period. The solids concentration of the biosolids is 3.1%. The surface area of the plate is 100 sq ft. If the downtime for biosolids cake discharge is 20 min, what is the net filter yield in pounds per hour per square foot?

Solution:
First, calculate solids loading rate, then multiply that number by the corrected time factor:

$$\text{Solids Loading Rate} = \frac{(\text{Biosolids, gph})(8.34 \text{ lb/gal})(\% \text{ Sol.}/100)}{\text{Plate Area, sq ft}}$$

$$= \frac{(300 \text{ gph})(8.34 \text{ lb/gal})(3.1/100)}{100 \text{ sq ft}}$$

$$= 0.78 \text{ lb/hr/sq ft}$$

Next, calculate net filter yield, using the corrected time factor:

$$\text{Net Filter Yield, lb/hr/sq ft} = \frac{(0.78\ \text{lb/hr/sq ft})\ (2\ \text{hr})}{2.33\ \text{hr}}$$

$$= 0.67\ \text{lb/hr/sq ft}$$

Problem 12.3

A 6 ft wide belt press receives a flow of 100 gpm of primary biosolids. What is the hydraulic loading rate in gallons per minute per foot?

Solution:

$$\text{Hydraulic Loading Rate, gpm/ft} = \frac{\text{Flow, gpm}}{\text{Belt Width, ft}}$$

$$= \frac{100\ \text{gpm}}{6\ \text{ft}}$$

$$= 16.7\ \text{gpm/ft}$$

Problem 12.4

A belt filter press 5 ft wide receives a primary biosolids flow of 140 gpm. What is the hydraulic loading rate in gallons per minute per square foot?

Solution:

$$\text{Hydraulic Loading Rate, gpm/ft} = \frac{\text{Flow, gpm}}{\text{Belt Width, ft}}$$

$$= \frac{140\ \text{gpm}}{5\ \text{ft}}$$

$$= 28\ \text{gpm/ft}$$

Problem 12.5

The amount of biosolids to be dewatered by the belt filter press is 20,000 lb/day. If the belt filter press is to be operated 10 hr each day, what should the biosolids feed rate in pounds per hour be to the press?

Solution:

$$\text{Biosolids Feed Rate, lb/hr} = \frac{\text{Biosolids to be dewatered, lb/day}}{\text{Operating Time, hr/day}}$$

$$= \frac{20{,}000\ \text{lb/day}}{10\ \text{hr/day}}$$

$$= 2000\ \text{lb/hr}$$

13 Wastewater Treatment Practice Calculations

13.1 An empty screenings hopper 4.3 ft by 5.8 ft is filled to an even depth of 28 in over the course of 96 hr. If the average plant flow rate was 4.9 MGD during this period, how many cubic feet of screenings were removed per million gallons of wastewater received?

13.2 A grit channel has a water depth of 1.4 ft and width of 1.7 ft. The flow rate through the channel is 700 gpm. What is the velocity through the channel in foot per second?

13.3 A grit channel has a water depth of 16 in and a width of 18 in. The flow rate through the channel is 1.2 MGD. What is the velocity through the channel in foot per second?

13.4 What is the gallon capacity of a wet well 12 ft long, 10 ft wide, and 6 ft deep?

DOI: 10.1201/9781003354314-13

13.5 A wet well is 14 ft long, 12 ft wide and contains water to a depth of 6 ft. How many gallons of water does it contain?

13.6 What is the cubic feet capacity of a wet well 9 ft by 9 ft with a maximum depth of 6 ft?

13.7 The maximum capacity of a wet well is 4,850 gal. If the wet well is 12 ft long and 8 ft wide, what is the maximum depth of water in the wet well?

13.8 A wet well is 10 ft long and 8 ft wide. If the wet well contains water to a depth of 3.1 ft, what is the volume of water in the wet well in gallons?

13.9 A wet well is 10 ft by 10 ft. During a 5 min pumping test, with no influent to the well, a pump lowers the water level 1.8 ft. What is the pumping rate in gallons per minute?

13.10 A wet well is 12 ft by 12 ft. During a 3 min pumping test, with no influent inflow, a pump lowers the water level 1.3 ft. What is the gallons per minute pumping rate?

13.11 The water level in a wet well drops 17 in during a 3 min pumping test. There was no influent to the wet well during the pumping test. If the wet well is 9 ft by 7 ft, what is the pumping rate in gallons per minute?

13.12 During a period when there is no pumping from the wet well, the water level rises 0.9 ft in 1 min. If the wet well is 9 ft long and 8 ft wide, what is the gallons per minute flow rate of wastewater entering the wet well?

13.13 A lift station wet well is 12 ft by 14 ft. For 5 min, the influent valve is closed and the well level drops 2.2 ft. What is the pumping rate in gallons per minute?

13.14 The influent valve to a 12 ft by 14 ft station wet well is closed for 4 min. During this time, the well level dropped 1.9 ft. What is the pump discharge in gallons per minute?

13.15 The dimensions of the wet well for a lift station are 10 ft, 9 in. by 12 ft, 2 in. The influent valve to the well is closed only long enough for the level to drop 2 ft. The time to accomplish this was 5 min and 30 sec. At what rate, in gallons per minute, is the pump discharging?

13.16 A lift station wet well is 11.8 ft by 14 ft. The influent flow to this well is 410 gpm. If the well level rises 1 in in 8 min, how many gallons per minute is the pump discharging?

13.17 A lift station wet well is 140 in by 148 in. The influent flow to this well is 430 gpm. If the well level drops 1.5 in in 5 min, how many gallons per minute is the pump discharging?

13.18 A lift station wet well is 9.8 ft by 14 ft and has an influent rate of 800 gpm. The level in the well drops 8 in in 15 min, and two pumps are in operation. If the first pump discharges at a rate of 500 gpm, at what pumping rate is the second pump discharging?

13.19 A total of 60 gal of screenings is removed from the wastewater flow during a 24 hr period. What is the screenings removal reported as cubic foot per day?

13.20 For 1 week a total 282 gal of screenings were removed from the wastewater screens. What was the average screenings removal in cubic foot per day?

13.21 The flow at a treatment plant is 3.33 MGD. If a total of 4.9 cu ft of screenings is removed during a 24 hr period, what is the screenings removal reported as cubic feet per million gallon?

13.22 On a particular day, a treatment plant receives a flow of 4.9 MGD. If 81 gal of screenings are removed that day, what is the screenings removal expressed as cubic feet per million gallon?

13.23 A total of 48 gal of screenings is removed from the treatment plant during a 24 hr period. If the treatment plant received a flow of 2,280,000 gpd, what is the screenings removal expressed as cubic feet per million gallon?

13.24 A screenings pit has a capacity of 600 cu ft. If an average of 2.9 cu ft of screenings is removed daily from the wastewater flow, in how many days will the pit be full?

13.25 A screenings pit has a capacity of 9 cu yds available for screenings. If the plant removes an average of 1.6 cu ft of screenings per day, in how many days will the pit be filled?

13.26 A plant has been averaging a screenings removal of 2.6 cu ft/MG. If the average daily flow is 2.9 MGD, how many days will it take to fill a screenings pit with an available capacity of 292 cu ft?

13.27 Suppose you want to use a screenings pit for 120 days. If the screenings removal rate is 3.5 cu ft/day, what is the required screenings pit capacity in cubic foot?

13.28 A grit channel is 4 ft wide, with water flowing to a depth of 18 in. If the flow meter indicates a flow rate of 1,820 gpm, what is the velocity of flow through the channel in foot per second?

13.29 A stick in a grit channel travels 26 ft in 32 sec. What is the estimated velocity in the channel in foot per second?

13.30 The total flow through both channels of a grit channel is 4.3 cfs. If each channel is 3 ft wide and water is flowing to a depth of 14 in, what is the velocity of flow through the channel in foot per second?

13.31 A stick is placed in a grit channel and flows 36 ft in 32 sec. What is the estimated velocity in the channel in foot per second?

13.32 The depth of water in a grit channel is 16 in. The channel is 34 in wide. If the flow meter indicates a flow of 1,140 gpm, what is the velocity of flow through the channel in foot per second?

13.33 A treatment plant removes 12 cu ft of grit in 1 day. If the plant flow is 8 MGD, what is this removal expressed as cubic feet per million gallon?

13.34 The total daily grit removal for a plant is 260 gal. If the plant flow is 11.4 MGD, how many cubic feet of grit are removed per MG flow?

13.35 The average grit removal at a particular treatment plant is 3.1 cu ft/MG. If the monthly average daily flow is 3.8 MGD, how many cubic yards of grit would be removed from the wastewater flow for one 30-day month?

13.36 The monthly average grit removal is 2.2 cu ft/MG. If the average daily flow for the month is 4,230,000 gpd, how many cubic yards must be available for grit disposal if the disposal pit is to have a 90-day capacity?

13.37 A grit channel 2.6 ft wide has water flowing to a depth of 16 in. If the velocity through the channel is 1.1 fps, what is the cubic feet per second flow rate through the channel?

13.38 A grit channel 3 ft wide has water flowing at a velocity of 1.4 fps. If the depth of the water is 14 in, what is the gallons per day flow rate through the channel?

13.39 A grit channel 32 in wide has water flowing to a depth of 10 in. If the velocity of the water is 0.90 fps, what is the cubic feet per second flow in the channel?

13.40 A suspended solids test was done on a 50 mL sample. The weight of the crucible and filter before the test was 25.6662 g. After the sample was filtered and dried, the cooled crucible weight was 25.6782 g. What was the concentration of suspended solids in milligrams per liter?

13.41 A 26.2345 g crucible was used to filter 26 mL of raw influent sample for a suspended solids test. The dried crucible weighed 26.2410 g. What was the concentration of suspended solids in milligrams per liter?

13.42 A BOD test was done on a 6 mL sample. The initial DO of the sample and dilution water was 8.42 mg/L. The DO of the sample after 5 days of incubation was 4.28 mg/L. What was the BOD of the sample?

13.43 A BOD test was done on a 5 mL sample. The initial DO of the sample and dilution water was 7.96 mg/L. The DO of the sample after 5 days of incubation was 4.26 mg/L. What was the BOD of the sample?

13.44 A wastewater treatment plant receives a flow of 3.13 MGD, with a total phosphorus concentration of 14.6 mg/L. How many pounds per day?

13.45 Raw influent BOD is 310 mg/L. If the influent flow rate is 6.15 MGD, at what rate are the pounds of BOD entering the plant?

13.46 The plant's influent flow rate of 4.85 MGD has a suspended solids concentration of 188 mg/L. How many pounds of suspended solids enter daily?

13.47 A plant has been averaging a screenings removal of 2.6 cu ft/MG. If the average daily flow is 2,950,000 gpd, how many days will it take to fill a screenings pit that has an available capacity of 270 cu ft?

13.48 For 7 days, a total of 210 gal of screenings was removed from the wastewater screens. What was the average screenings removal in cubic foot per day?

13.49 A total of 5.4 cu ft of screenings is removed from the wastewater flow during a 24 hr period. If the flow at the treatment plant is 2,910,000 gpd, what is the screenings removal reported as cubic feet per million gallon?

13.50 A screenings pit has a capacity of 12 cu yd available for screenings. If the plant removes an average of 2.4 cu ft of screenings per day, in how many days will the pit be filled?

13.51 A float is placed in a channel. If the float travels 36 ft in 31 sec, what is the estimated velocity in the channel in foot per second?

13.52 A grit channel 2.6 ft wide has water flowing to a depth of 15 in. If the velocity of the water is 0.8 fps, what is the cubic feet per second flow in the channel?

13.53 The total daily grit removal for a treatment plant is 210 gal. If the plant flow is 8.8 MGD, how many cubic feet of grit are removed per MG flow?

13.54 A grit channel is 2.6 ft wide, with water flowing to a depth of 15 in. If the flow velocity through the channel is 1.8 ft/sec, what is the gallons per minute flow through the clarifier?

13.55 The average grit removal at a particular treatment plant is 2.3 cu ft/MG. If the monthly average daily flow is 3,610,000 gpd, how many cubic yards of grit are expected to be removed from the wastewater flow during a 30-day month?

13.56 A grit channel 3 ft wide has water flowing to a depth of 10 in. If the velocity through the channel is 1 fps, what is the cubic feet per second flow rate through the channel?

13.57 A circular clarifier has a capacity of 160,000 gal. If the flow through the clarifier is 1,810,000 gpd, what is the detention time in hours for the clarifier?

13.58 A flow to a sedimentation tank 90 ft long, 25 ft wide, and 10 ft deep is 3.25 MGD. What is the detention time in the tank in hours?

13.59 A circular clarifier receives a steady, continuous flow of 4,350,000 gpd. If the clarifier is 90 ft in diameter and 12 ft deep, what is the clarifier detention time in hours?

13.60 The influent flow rate to a primary settling tank is 2.01 MGD. The tank is 84 ft in length, 20 ft wide and has a water depth of 13.1 ft. What is the detention time of the tank in hours?

13.61 A primary settling tank 90 ft long, 20 ft wide, and 14 ft deep receives a flow rate of 1.45 MGD. What is the surface overflow rate in gallons per day per square foot?

13.62 A primary sludge sample is tested for total solids. The dish alone weighed 22.20 g. The sample with the dish weighed 73.86 g. After drying, the dish with dry solids weighed 23.10 g. What was the percent total solids (% TS) of the sample?

13.63 Primary sludge is pumped to a gravity thickener at 390 gpm. The sludge concentration is 0.8%. How many pounds of solids are pumped daily?

13.64 The raw influent suspended solids concentration is 140 mg/L. The primary effluent concentration of suspended solids is 50 mg/L. What percentage of the suspended solids is removed by primary treatment?

13.65 A primary tank with a total weir length of 80 ft receives a flow rate of 1.42 MGD. What is the weir overflow rate in gallons per day per foot?

13.66 A wastewater treatment plant has 8 primary tanks. Each tank is 80 ft long, 20 ft wide, with a side water depth of 12 ft and a total weir length of 86 ft. The flow rate to the plant is 5 MGD. There are three tanks currently in service. Calculate the detention time in minutes, the surface overflow rate in gallons per day per square foot, and the weir overflow rate in gallons per day per foot.

13.67 The flow to a sedimentation tank 80 ft long, 35 ft wide, and 12 ft deep is 3.24 MGD. What is the detention time in hours in the tank?

13.68 A rectangular clarifier has a total of 112 ft of weir. What is the weir overflow rate in gallons per day per foot when the flow is 1,520,000 gpd?

13.69 A circular clarifier receives a flow of 2.98 MGD. If the diameter of the weir is 70 ft, what is the weir overflow rate in gallons per day per foot?

13.70 The average flow to a clarifier is 2,520 gpm. If the diameter of the wire is 90 ft, what is the weir overflow rate in gallons per day per foot?

13.71 The total feet of weir for a clarifier is 192 ft. If the flow to the weir is 1.88 MGD, what is the weir overflow rate in gallons per day per foot?

13.72 A circular clarifier has a diameter of 70 ft. If the primary clarifier influent flow is 2,910,000 gpd, what is the surface overflow rate in gallons per day per square foot?

13.73 A sedimentation basin 80 ft by 30 ft receives a flow of 2.35 MGD. What is the surface overflow rate in gallons per day per square foot?

13.74 A sedimentation tank is 80 ft long and 30 ft wide. If the flow to the tank is 2,620,000 gpd, what is the surface overflow rate in gallons per day per square foot?

13.75 The average flow to a secondary clarifier is 2,610 gpm. What is the surface overflow rate in gallons per day per square foot if the secondary clarifier has a diameter of 60 ft?

13.76 A secondary clarifier is 70 ft in diameter and receives a combined primary effluent and return activated sludge (RAS) flow of 4.1 MGD. If the MLSS concentration in the aeration tank is 3,110 mg/L, what is the solids loading rate on the secondary clarifier in pounds per day per square foot?

13.77 A secondary clarifier, 80 ft in diameter, receives a primary effluent flow of 3.3 MGD and a return sludge flow of 1.1 MGD. If the MLSS concentration is 3,220 mg/L, what is the solids loading rate on the clarifier in pounds per day per square foot?

13.78 The MLSS concentration in an aeration tank is 2,710 mg/L. The 70 ft diameter secondary clarifier receives a combined primary effluent and return activated sludge (RAS) flow of 3,220,000 gpd. What is the solids loading rate on the secondary clarifier in pound SS per day per square foot?

13.79 A secondary clarifier, 80 ft in diameter, receives a primary effluent flow of 2,320,000 gpd and a return sludge flow of 660,000 gpd. If the MLSS concentration is 3,310 mg/L, what is the solids loading rate on the clarifier in pounds per day per square foot?

13.80 If 125 mg/L suspended solids are removed by a primary clarifier, how many pounds per day suspended solids are removed when the flow is 5,550,000 gpd?

13.81 The flow to a primary clarifier is 2,920,000 gpd. If the influent to the clarifier has a suspended solids concentration of 240 mg/L and the primary effluent has 200 mg/L SS, how many pounds per day suspended solids are removed by the clarifier?

13.82 The flow to a secondary clarifier is 4.44 MGD. If the influent BOD concentration is 200 mg/L and the effluent concentration is 110 mg/L, how many pounds of BOD are removed daily?

13.83 The flow to a primary clarifier is 980,000 gpd. If the influent to the clarifier has suspended solids concentration of 320 mg/L and the primary effluent has a suspended solids concentration of 120 mg/L, how many pounds per day suspended solids are removed by the clarifier?

13.84 The suspended solids entering a primary clarifier are 230 mg/L. If the suspended solids in the primary clarifier effluent are 95 mg/L, what is the suspended solids removal efficiency of the primary clarifier?

13.85 The suspended solids entering a primary clarifier are 188 mg/L. If the suspended solids in the primary clarifier effluent is 77 mg/L, what is the suspended solids removal efficiency of the primary clarifier?

13.86 The influent to a primary clarifier has a BOD content of 280 mg/L. If the primary clarifier effluent has a BOD concentration of 60 mg/L, what is the BOD removal efficiency of the primary clarifier?

13.87 The BOD concentration of a primary clarifier is 300 mg/L. If the primary clarifier effluent BOD concentration is 189 mg/L, what is the BOD removal efficiency of the primary clarifier?

13.88 The flow to a circular clarifier is 4,120,000 gpd. If the clarifier is 80 ft in diameter and 10 ft deep, what is the clarifier detention time in hours?

13.89 A circular clarifier has a diameter of 60 ft. If the primary clarifier influent flow is 2,320,000 gpd, what is the surface overflow rate in gallons per day per square foot?

13.90 A rectangular clarifier has a total of 215 ft of weir. What is the weir overflow rate in gallons per day per foot when the flow is 3,728,000 gpd?

13.91 A secondary clarifier, 60 ft in diameter, receives a primary effluent flow of 1,910,000 gpd and a return sludge flow of 550,000 gpd. If the MLSS concentration is 2,710 mg/L, what is the solids loading rate in pounds per day per square foot on the clarifier?

13.92 A circular primary clarifier has a diameter of 70 ft. If the influent flow to the clarifier is 3.10 MGD, what is the surface overflow rate in gallons per day per square foot?

13.93 A secondary clarifier, 80 ft in diameter, receives a primary effluent flow of 3,150,000 gpd and a return sludge flow of 810,000 gpd. If the mixed liquor suspended solids concentration is 2,910 mg/L, what is the solids loading rate in the clarifier in pounds per day per square foot?

13.94 The flow to a secondary clarifier is 5.3 MGD. If the influent BOD concentration is 228 mg/L and the effluent BOD concentration is 110 mg/L, how many pounds per day BOD are removed daily?

13.95 The flow to a sedimentation tank 90 ft, 40 ft wide, and 14 ft deep is 5.10 MGD. What is the detention time in the tank in hours?

13.96 The average flow to a clarifier is 1,940 gpm. If the diameter of the wire is 70 ft, what is the weir overflow rate in gallons per day per foot?

13.97 The flow to a secondary clarifier is 4,440,000 gpd. How many pounds of BOD are removed daily if the influent BOD concentration is 190 mg/L and the effluent BOD concentration is 106 mg/L?

13.98 The flow to a primary clarifier is 3.88 MGD. If the influent to the clarifier has suspended solids concentration of 290 mg/L and the primary clarifier effluent has a suspended solids concentration of 80 mg/L, how may pounds per day suspended solids are removed by the clarifier?

13.99 The primary clarifier influent has a BOD concentration of 260 mg/L. If the primary clarifier effluent has a BOD concentration of 69 mg/L, what is the BOD removal efficiency of the primary clarifier?

13.100 A sedimentation tank is 90 ft long and 40 ft wide. If the flow to the tank is 2,220,000 gpd, what is the surface overflow rate in gallons per day per square foot?

13.101 A trickling filter, 80 ft in diameter, treats a primary effluent flow of 660,000 gpd. If the recirculated flow to the trickling filter is 120,000 gpd, what is the hydraulic loading rate in gallons per day per square foot on the trickling filter?

13.102 A high-rate trickling filter receives a flow of 2,360 gpm. If the filter has a diameter of 90 ft, what is the hydraulic loading rate in gallons per day per square foot on the filter?

13.103 The total influent flow (including recirculation) to a trickling filter is 1.5 MGD. If the trickling filter is 90 ft in diameter, what is the hydraulic loading rate in gallons per day per square foot on the trickling filter?

13.104 A high-rate trickling filter receives a daily flow of 2.1 MGD. What is the hydraulic loading rate in million gallons per day per acre if the filter is 96 ft in diameter?

13.105 A trickling filter 100 ft in diameter, with a media depth of 6 ft, receives a flow of 1,400,000 gpd. If the BOD concentration of the primary effluent is 210 mg/L, what is the organic loading on the trickling filter in pounds BOD per day per 1,000 cu ft?

13.106 A 90 ft diameter trickling filter with a media depth of 7 ft receives a primary effluent flow of 3,400,000 gpd with a BOD of 111 mg/L. What is the organic loading on the trickling filter in pounds BOD per day per 1,000 cu ft?

13.107 A trickling filter, 80 ft in diameter, with a media dept of 7 ft, receives a flow of 0.9 MGD. If the BOD concentration of the primary effluent is 201 mg/L, what is the organic loading on the trickling filter in pounds BOD per day per 1,000 cu ft?

13.108 A trickling filter has a diameter of 90 ft and a media depth of 5 ft. The primary effluent has a BOD concentration of 120 mg/L. If the total flow to the filter is 1.4 MGD, what is the organic loading in pounds per acre-foot?

13.109 If 122 mg/L suspended solids are removed by a trickling filter, how many pounds per day suspended solids are removed when the flow is 3,240,000 gpd?

13.110 The flow to a trickling filter is 1.82 MGD. If the primary effluent has a BOD concentration of 250 mg/L and the trickling filter effluent has a BOD concentration of 74 mg/L, how many pounds of BOD are removed?

13.111 If 182 mg/L of BOD are removed from a trickling filter when the flow to the trickling filter is 2,920,000 gpd, how many pounds per day BOD are removed?

13.112 The flow to a trickling filter is 5.4 MGD. If the trickling filter effluent has a BOD concentration of 28 mg/L and the primary effluent has a BOD concentration of 222 mg/L, how many pounds of BOD are removed daily?

13.113 The suspended solids concentration entering a trickling filter is 149 mg/L. If the suspended solids concentration in the trickling filter effluent is 48 mg/L, what is the suspended solids removal efficiency of the trickling filter?

13.114 The influent to a primary clarifier has a BOD content of 261 mg/L. The trickling filter effluent BOD is 22 mg/L. What is the BOD removal efficiency of the treatment plant?

13.115 The suspended solids entering a trickling filter are 201 mg/L. If the suspended solids in the trickling filter effluent are 22 mg/L, what is the suspended solids removal efficiency of the trickling filter?

13.116 The suspended solids concentration entering a trickling filter is 111 mg/L. If 88 mg/L suspended solids are removed from the trickling filter, what is the suspended solids removal efficiency of the trickling filter?

13.117 A treatment plant receives a flow of 3.4 MGD. If the trickling filter effluent is recirculated at the rate of 3.5 MGD, what is the recirculation ratio?

13.118 The influent to the trickling filter is 1.64 MGD. If the recirculated flow is 2.32 MGD, what is the recirculation ratio?

13.119 The trickling filter effluent is recirculated at the rate of 3.86 MGD. If the treatment plant receives a flow of 2.71 MGD, what is the recirculation ratio?

13.120 A trickling filter has a desired recirculation ratio of 1.6. If the primary effluent flow is 4.6 MGD, what is the desired recirculated flow in million gallons per day?

13.121 A trickling filter 90 ft in diameter treats a primary effluent flow rate of 0.310 MGD. If the recirculated flow to the clarifier is 0.355 MGD, what is the hydraulic loading rate on the trickling filter in gallons per day per square foot (gpd/ft^2)?

13.122 A treatment plant receives a flow rate of 2.8 MGD. If the trickling filter effluent is recirculated at a rate of 4.55 MGD, what is the recirculated ratio?

13.123 A trickling filter 80 ft in diameter with a media depth of 6 ft receives a primary effluent flow rate of 1,350,000 gpd. If the population equivalent BOD is 75 mg/L, what is the organic loading rate on the unit in pounds per day per 1,000 cubic feet (lb/day/100 cu ft)?

13.124 The flow rate to a trickling filter is 4.1 MGD. If the population equivalent BOD is 81 mg/L and the secondary effluent BOD is 13 mg/L, how many pounds of BOD are removed daily?

13.125 A standard-rate filter, 80 ft in diameter, treats a primary effluent flow of 520,000 gpd. If the recirculated flow to the trickling filter is 110,000 gpd, what is the hydraulic loading rate on the filter in gallons per day per square foot?

13.126 If 114 mg/L suspended solids are removed by a trickling filter, how many pounds per day suspended solids are removed when the flow is 2,840,000 gpd?

13.127 The suspended solids concentration entering a trickling filter is 200 mg/L. If the suspended solids concentration in the trickling filter effluent is 69 mg/L, what is the suspended solids removal efficiency of the trickling filter?

13.128 The flow to a trickling filter is 1.44 MGD. If the primary effluent has a BOD concentration 242 mg/L and the trickling filter effluent has a BOD concentration of 86 mg/L, how many pounds of BOD are removed?

13.129 A high-rate trickling filter receives a combined primary effluent and recirculated flow of 2.88 MGD. If the filter has a diameter of 90 ft, what is the hydraulic loading rate on the filter in gallons per day per square foot?

13.130 The influent of a primary clarifier has a BOD content of 210 mg/L. The trickling filter effluent BOD is 22 mg/L. What is the BOD removal efficiency of the treatment plant?

13.131 A trickling filter has a diameter of 90 ft and an average media depth of 6 ft. The primary effluent has a BOD concentration of 144 mg/L. If the total flow to the filter is 1.26 MGD, what is the organic loading in pounds per day per acre-foot?

13.132 The flow to a trickling filter is 4.22 MGD. If the trickling filter effluent has a BOD concentration of 21 mg/L and the primary effluent has a BOD concentration of 199 mg/L, how many pounds of BOD are removed daily?

13.133 A treatment plant receives a flow of 3.6 MGD. If the trickling filter effluent is recirculated at the rate of 3.8 MGD, what is the recirculation ratio?

13.134 A high-rate trickling filter receives a daily flow of 1.9 MGD. What is the hydraulic loading rate in million gallons per day per acre if the filter is 80 ft in diameter?

13.135 The total influent flow (including recirculation) to a trickling filter is 1.93 MGD. If the trickling filter is 90 ft diameter, what is the hydraulic loading in gallons per day per square foot on the trickling filter?

13.136 A trickling filter 70 ft in diameter, with a media depth of 6 ft, receives a flow of 0.81 MGD. If the BOD concentration of the primary effluent is 166 mg/L, what is the organic loading on the trickling filter in pounds BOD per day per 1,000 cu ft?

13.137 The influent to the trickling filter is 1.67 MGD. If the recirculated flow is 2.35 MGD, what is the recirculation ratio?

13.138 The suspended solids concentration entering a trickling filter is 243 mg/L. If the suspended solids concentration of the trickling filter effluent is 35 mg/L, what is the suspended solids removal efficiency of the trickling filter?

13.139 A rotating biological contactor (RBC) treats a primary effluent flow of 2.98 MGD. If the media surface area is 720,000 MGD sq ft, what is the hydraulic loading rate in gallons per day per square foot on the RBC?

13.140 A rotating biological contactor treats a flow of 4,725,000 gpd. The manufacturer data indicates a media surface area of 880,000 sq ft. What is the hydraulic loading rate in gallons per day per square foot on the RBC?

13.141 Data indicates a media surface area of 440,000 sq ft. A rotating biological contactor treats a flow of 1.55 MGD. What is the hydraulic loading rate in gallons per day per square foot on the RBC?

13.142 A rotating biological contactor has a media area of 800,000 sq ft. For a maximum hydraulic loading of 7 gpd/sq ft, what is the desired gallons per day flow to the contactor?

13.143 The suspended solids concentration of a wastewater is 241 mg/L. If the normal K-value at the plant is 0.55, what is the estimated particulate BOD concentration of the wastewater?

13.144 The wastewater entering a rotating biological contactor has a BOD content of 222 mg/L. The suspended solids content is 241 mg/L. If the K-value is 0.5, what is the estimated soluble BOD (in milligrams per liter) of the wastewater?

13.145 The wastewater entering a rotating biological contactor has a BOD content of 240 mg/L. The suspended solids concentration of the wastewater is 150 mg/L. If the K-value is 0.5, what is the estimated soluble BOD (in milligrams per liter) of the wastewater?

13.146 A rotating biological contactor receives a 1.9 MGD flow, with a BOD concentration of 288 mg/L and an SS concentration of 268 mg/L. If the K-value is 0.6, how many pounds per day soluble BOD enter the RBC?

13.147 A rotating biological contactor (RBC) receives a flow of 2.8 MGD. The BOD of the influent wastewater to the RBC is 187 mg/L, and the surface area of the media is 765,000 sq ft. If the suspended solids concentration of the wastewater is 144 mg/L and the K-value is 0.52, what is the organic loading rate in pounds per day per 1,000 sq ft?

13.148 An RBC unit treats a flow rate of 0.45 MGD. The two shafts used provide a total surface area of 190,000 sq ft. What is the hydraulic loading on the unit in gallons per day per square foot?

13.149 The influent to an RBC has a total BOD concentration of 190 mg/L and a suspended solids concentration of 210 mg/L. If there are 0.6 lb of particulate BOD per pound of suspended solids, estimate the soluble BOD concentration in milligrams per liter.

13.150 An RBC receives a flow rate of 1.9 MGD. If the influent soluble BOD concentration is 128 mg/L and the total media surface area is 410,000 sq ft for the RBC unit, what is the organic loading in pounds per day per 1,000 sq ft?

13.151 Estimate the soluble BOD loading on an RBC treating a flow rate of 0.71 MGD. The total unit surface area is 110,000 sq ft. The total BOD concentration is 210 mg/L, with a suspended solids concentration of 240 mg/L and a K-value of 0.65.

13.152 An RBC unit contains two shafts operated in series each with a surface area of 103,000 sq ft. The shafts can both be partitioned by baffles at 25% shaft length intervals. Currently, the first stage of the RBC unit is baffled to use 75% of one of the two shafts. The unit receives a flow rate of 0.455 MGD. The primary effluent total BOD concentration is 241 mg/L. The suspended solids concentration is 149 mg/L, and the value of K is 0.5. Calculate the hydraulic loading, unit organic loading, and first-stage organic loading.

13.153 An RBC treats a primary effluent flow of 2.96 MGD. If the media surface area is 660,000 sq ft, what is the hydraulic loading rate in gallons per day per square foot on the RBC?

13.154 The suspended solids concentration of a wastewater is 222 mg/L. If the normal K-value at the plant is 0.5, what is the estimated particulate BOD concentration of the wastewater?

13.155 An RBC receives a 2.9 MGD flow with a BOD concentration of 205 mg/L and SS of 210 mg/L. If the K-value is 0.6, how many pounds per day soluble BOD enter the RBC?

13.156 The wastewater flow to an RBC is 2,415,000 gpd. The wastewater has a soluble BOD concentration of 121 mg/L. The RBC media has total of 760,000 sq ft. What is the organic loading rate in pounds per day per 1,000 sq ft on the RBC?

13.157 An aeration tank is 80 ft long, 25 ft wide and operates at an average depth of 14 ft. What is the capacity of the tank in gallons?

13.158 What is the gallon capacity of an aeration tank that is 80 ft log, 30 ft wide and operates at an average depth of 12 ft?

13.159 A secondary clarifier has a diameter of 80 ft and an average depth of 12 ft. What is the volume of water in the clarifier in gallons?

13.160 A clarifier has a diameter of 70 ft and an average depth of 10 ft. What is the volume of water in the clarifier in gallons?

13.161 The flow to an aeration tank is 880,000 gpd. If the BOD content of the wastewater entering the aeration tank is 240 mg/L, what is the pounds per day BOD loading?

13.162 The flow to an aeration tank is 2,980 gpm. If the COD concentration of the wastewater is 160 mg/L, how many pounds of COD are applied to the aeration tank daily?

13.163 The BOD content of the wastewater entering an aeration tank is 165 mg/L. If the flow to the aeration tank is 3,240,000 gpd, what is the pounds per day BOD loading?

13.164 The daily flow to an aeration tank is 4,880,000 gpd. If the COD concentration of the influent wastewater is 150 mg/L, how many pounds of COD are applied to the aeration tank daily?

13.165 If the mixed liquor suspended solids concentration is 2,110 mg/L and the aeration tank has a volume of 460,000 gal, how many pounds of suspended solids are in the aeration tank?

13.166 The aeration tank of a conventional activated sludge plant has a mixed liquor volatile suspended solids concentration of 2,420 mg/L. If the aeration tank is 90 ft long, 50 ft wide and has wastewater to a depth of 16 ft, how many pounds of MLVSS are under aeration?

13.167 The aeration tank of a conventional activated sludge plant has a mixed liquor volatile suspended solids concentration of 2,410 mg/L. If the aeration tank is 80 ft long, 40 ft wide and has wastewater to a depth of 16 ft, how many pounds of MLVSS are under aeration?

13.168 The aeration tank is 110 ft long, 30 ft wide, and has wastewater to a depth of 16 ft. If the aeration tank of this conventional activated sludge plant has mixed liquor suspended solids concentration of 2,740 mg/L, how many pounds of MLSS are under aeration?

13.169 An aeration basin is 110 ft long, 50 ft wide, and has wastewater to a depth of 16 ft. If the mixed liquor suspended solids concentration in the aeration tank is 2,470 mg/L, with a volatile solids content of 73%, how many pounds of MLVSS are under aeration?

13.170 An activated sludge aeration tank receives a primary effluent flow of 2.72 MGD, with a BOD concentration of 198 mg/L. The mixed liquor volatile suspended solids concentration is 2,610 mg/L, and the aeration tank volume is 480,000 gal. What is the current F/M ratio?

13.171 An activated sludge aeration tank receives a primary effluent flow of 3,350,000 gpd with a BOD of 148 mg/L. The mixed liquor volatile suspended solids are 2,510 mg/L, and the aeration tank volume is 490,000 gal. What is the current F/M ratio?

13.172 The flow to a 195,000 gal oxidation ditch is 320,000 gpd. The BOD concentration of the wastewater is 180 mg/L. If the mixed liquor suspended solids concentration is 2,540 mg/L with a volatile solids content of 72%, what is the F/M ratio?

13.173 The desired F/M ratio at an extended aeration activated sludge plant is 0.7 lb BOD/lb MLVSS. If the 3.3 MGD primary effluent flow has a BOD of 181 mg/L, how many pounds of MLVSS should be maintained in the aeration tank?

13.174 The desired F/M ratio at a particular activated sludge plant is 0.4 lb BOD/lb MLVSS. If the 2,510,000 gpd primary effluent flow has a BOD concentration of 141 mg/L, how may pounds of MLVSS should be maintained in the aeration tank?

13.175 An aeration tank has a total of 16,100 lb of mixed liquor suspended solids. If a total of 2,630 lb/day suspended solids enters the aeration tank in the primary effluent flow, what is the sludge age in the aeration tank?

13.176 An aeration tank contains 480,000 gal of wastewater, with an MLSS concentration of 2,720 mg/L. If the primary effluent flow is 2.9 MGD, with a suspended solids concentration of 110 mg/L, what is the sludge age?

13.177 An aeration tank is 110 ft long, 50 ft wide and operates at a depth of 14 ft. The MLSS concentration in the aeration tank is 2,510 mg/L. If the influent flow to the tank is 2.88 MGD, with a suspended solids concentration of 111 mg/L, what is the sludge age?

13.178 The MLSS concentration in the aeration tank is 2,960 mg/L. The aeration tank is 110 ft long, 50 ft wide and operates at a depth of 14 ft. If the influent flow to the tank is 1.98 MGD and has a suspended solids concentration of 110 mg/L, what is the sludge age?

13.179 An oxidation ditch has a volume of 211,000 gal. The 270,000 gpd flow to the oxidation ditch has a suspended solids concentration of 205 mg/L. If the MLSS concentration is 3,810 mg/L, what is the sludge age in the oxidation ditch?

13.180 An activated sludge system has a total of 29,100 lb of mixed liquor suspended solids. The suspended solids leaving the final clarifier in the effluent are calculated to be 400 lb/day. The pounds suspended solids wasted from the final clarifier are 2,920 lb/day. What is the solids retention time in days?

13.181 Determine the solids retention time (SRT) given the following data:

Aeration tank volume: 1,500,000 gal
Final clarifier: 106,000 gal
Population flow: 3.3 MGD
WAS pumping rate: 72,000 gpd

MLSS: 2,710 mg/L
WAS: 5,870 mg/L
Sec. eff. SS: 25 mg/L
Ave. clarifier core SS: 1,940 mg/L

13.182 An aeration tank has a volume of 460,000 gal. The final clarifier has a volume of 178,000 gal. The MLSS concentration in the aeration tank is 2,222 mg/L. If a total of 1,610 lb/day SS is wasted and 240 lb/day SS are in the secondary effluent, what is the solids retention time for the activated sludge system?

13.183 Calculate the solids retention time (SRT) given the following data:

Aeration tank volume: 350,000 gal
Final clarifier: 125,000 gal
Population equivalent flow: 1.4 MGD
WAS pumping rate: 27,000 gpd

MLSS: 2,910 mg/L
WAS: 6,210 mg/L
Sec. eff. SS: 16 mg/L

13.184 The settleability test after 30 min indicates a sludge settling volume of 220 mL/L. Calculate the RAS flow as a ratio to the secondary influent flow.

13.185 A total of 280 mL/L sludge settled during a settleability test after 30 min. The secondary influent flow is 3.25 MGD. Calculate the RAS flow.

13.186 Given the following data, calculate the RAS return rate using the aeration tank solids balance equation.

MLSS: 2,200 mg/L
RAS SS: 7,520 mg/L
Population equivalent: 6.4 MGD

13.187 The desired F/M ratio for an activated sludge system is 0.5 lb BOD/lb MLVSS. It has been calculated that 3,400 lb of BOD enter the aeration tank daily. If the volatile solids content of the MLSS is 69%, how many pounds MLSS are desired in the aeration tank?

13.188 Using a desired sludge age, it was calculated that 14,900 lb MLSS are desired in the aeration tank. If the aeration tank volume is 790,000 gal and the MLSS concentration is 2,710 mg/L, how many pounds per day MLSS should be wasted?

13.189 Given the following data, determine the pounds per day SS to be wasted:

Aeration tank volume: 1,200,000 gal
Influent flow: 3,100,000 gpd
BOD: 110 mg/L
Desired F/M: 0.4 lb BOD/day/lb MLVSS
MLSS: 2,200 mg/L
% VS: 68%

13.190 The desired sludge age for an activated sludge plant is 5.6 days. The aeration tank volume is 910,000 gal. If 3,220 lb/day suspended solids enter the aeration tank and the MLSS concentration is 2,900 mg/L, how many pounds per day MLSS (suspended solids) should be wasted?

13.191 The desired SRT for an activated sludge plant is 9 days. There are a total of 32,400 lb SS in the system. The secondary effluent flow is 3,220,000 gpd, with a suspended solids content of 23 mg/L. How many pounds per day WAS SS must be wasted to maintain the desired SRT?

13.192 It has been determined that 5,580 lb/day solids must be removed from the secondary system. If the RAS SS concentration is 6,640 mg/L, what must be the WAS pumping rate in million gallons per day?

13.193 The WAS suspended solids concentration is 6,200 mg/L. If 8,710 lb/day dry solids are to be wasted, what must the WAS pumping rate be in million gallons per day?

13.194 Given the following data, calculate the WAS pumping rate required (in million gallons per day):

Desired SRT: 9 days
Clarifier + aerator volume: 1.8 MG
MLSS: 2,725 mg/L
RAS SS: 7,420 mg/L
Sec. eff. SS: 18 mg/L
Inf. flow: 4.3 MGD

13.195 Given the following data, calculate the WAS pumping rate required (in million gallons per day):

Desired SRT: 8.5 days
Clarifier + aerator volume: 1.7 MG
MLSS: 2,610 mg/L
RAS SS: 6,100 mg/L
Sec. eff. SS: 14 mg/L
Inf. flow: 3.8 MGD

13.196 An oxidation ditch has a volume of 166,000 gal. If the flow to the oxidation ditch is 190,000 gpd, what is the detention time in hours?

13.197 An oxidation ditch receives a flow of 0.23 MGD. If the volume of the oxidation ditch is 370,000 gal, what is the detention time in hours?

13.198 If the volume of the oxidation ditch is 420,000 gal and the oxidation ditch receives a flow of 305,000 gpd, what is the detention time in hours?

13.199 The volume of an oxidation ditch is 210,000 gal. If the oxidation ditch receives a flow of 310,000 gpd, what is the detention time in hours?

13.200 An aeration tank is 80 ft long, 40 ft wide and operates at an average depth of 15 ft. What is the capacity of the tank in gallons?

13.201 The BOD content of the wastewater entering an aeration tank is 220 mg/L. If the flow to the aeration tank is 1,720,000 gpd, what is the pounds per day BOD loading?

13.202 The flow to a 220,000 gal oxidation ditch is 399,000 gpd. The BOD concentration of the wastewater is 222 mg/L. If the mixed liquor suspended solids concentration is 3,340 mg/L, with a volatile solids content of 68%, what is the F/M ratio?

13.203 A clarifier has a diameter of 90 ft and an average depth of 12 ft. What is the capacity of the clarifier in gallons?

13.204 The daily flow to an aeration tank is 3,920,000 gpd. If the COD concentration of the influent wastewater is 160 mg/L, how many pounds of COD are applied to the aeration tank daily?

13.205 An aeration tank contains 530,000 gal of wastewater, with an MLSS concentration of 2,700 mg/L. If the primary effluent flow is 1.8 MGD, with a suspended solids concentration of 190 mg/L, what is the sludge age?

13.206 A mixed liquor sample is poured into a 2,100 mL settlometer. After 30 min, there is a settled sludge volume of 440 mL. If the plant flow rate (Q) is 6.1 MGD, what should the return sludge flow rate be in gallons per minute?

13.207 The mixed liquor is a 0.45 MG aeration tank that has a suspended solids concentration (MLSS) of 2,100 mg/L. The waste sludge is being removed at a rate of 0.120 MGD and has a concentration of 4,920 mg/L. If the target MLSS is 2,050 mg/L, what should the new waste sludge pumping rate be?

13.208 The mixed liquor in a 0.44 MG aeration tank has a suspended solids concentration (MLSS) of 2,090 mg/L. The waste sludge is being removed at a rate of 87.3 gpm and has a concentration of 4,870 mg/L. If the target MLSS is 2,170 mg/L, what should the new waste sludge pumping rate be in gallons per minute?

13.209 An aeration tank has an MLSS concentration of 2,210 mg/L. The volume of the tank is 0.66 MG. The plant flow rate is 3.25 MGD. The primary effluent suspended solids concentration is 131 mg/L. What is the sludge age?

13.210 The desired F/M ratio at a particular activated sludge plant is 0.6 lb BOD/lb MLVSS. If the 2.88 MGD primary effluent flow has a BOD concentration of 146 mg/L, how many pounds of MLVSS should be maintained in the aeration tank?

13.211 An oxidation ditch receives a flow of 0.31 MGD. If the volume of the oxidation ditch is 410,000 gal, what is the detention time in hours?

13.212 The desired F/M ratio at a particular activated sludge plant is 0.8 lb COD/lb MLVSS. If the 2,410,000 gpd primary effluent flow has a COD concentration of 161 mg/L, how many pounds MLVSS should be maintained in the aeration tank?

13.213 An aeration tank is 110 ft long, 40 ft wide and operates at a depth of 14 ft. The MLSS concentration in the aeration tank is 2,910 mg/L. If the influent flow to the tank is 1.4 MGD and contains a suspended solids concentration of 170 mg/L, what is the sludge age?

13.214 If the volume of the oxidation ditch is 620,000 gal and an oxidation ditch receives a flow of 0.36 MGD, what is the detention time in hours?

13.215 An oxidation ditch has a volume of 260,000 gal. The 0.4 MGD flow to the oxidation ditch has a suspended solids concentration of 200 mg/L. If the MLSS concentration is 3,980 mg/L, what is the sludge age in the oxidation ditch?

13.216 If the mixed liquor suspended solids concentration is 2,710 mg/L and the aeration tank has a volume of 440,000 gal, how many pounds of suspended solids are in the aeration tank?

13.217 The desired F/M ratio at a conventional activated sludge plant is 0.4 lb BOD/lb MLVSS. If the 2.88 MGD primary effluent flow has a BOD of 146 mg/L, how many pounds of MLVSS should be maintained in the aeration tank?

13.218 The aeration tank of a conventional activated sludge plant has a mixed liquor volatile suspended solids concentration of 2,510 mg/L. If the aeration tank is 110 ft long, 40 ft wide and has wastewater to a depth of 18 ft, how many pounds of MLVSS are in the aeration tank?

13.219 The MLSS concentration in an aeration tank is 2,740 mg/L. The aeration tank contains 710,000 gal of wastewater. If the primary effluent flow is 1.86 MGD, with a suspended solids concentration of 184 mg/L, what is the sludge age?

13.220 Determine the solids retention time (SRT) given the following data:

Aeration tank volume: 1,410,000 gal	MLSS: 2,680 mg/L
Final clarifier: 118,000 gal	WAS: 5,870 mg/L
Population equivalent flow: 3.1 MGD	Sec. Eff. SS: 20 mg/L
WAS: 76,000 gpd	Clarifier core SS: 1,910 mg/L

13.221 The settleability test after 30 min indicates a sludge settling volume of 231 mL/L. Calculate the RAS flow as a ratio to the secondary influent flow.

13.222 The desired F/M ratio at an activated sludge plant is 0.5 lb BOD/lb MLVSS. It was calculated that 3,720 lb/day BOD enter the aeration tank. If the volatile solids content of the MLSS is 70%, how many pounds MLSS are desired in the aeration tank?

13.223 The desired sludge age for a plant is 5 days. The aeration tank volume is 780,000 gal. If 3,740 lb/day suspended solids enter the aeration tank and the MLSS concentration is 2,810 mg/L, how many pounds per day MLSS should be wasted?

13.224 It has been determined that 4,110 lb/day of dry solids must be removed from the secondary system. If the RAS SS concentration is 6,410 mg/L, what must be the WAS pumping rate in million gallons per day?

13.225 Calculate the BOD loading (pounds per day) on a pond if the influent flow is 410,000 gpd, with a BOD of 250 mg/L.

13.226 The BOD concentration of the wastewater entering a pond is 161 mg/L. If the flow to the pond is 225,000 gpd, how many pounds per day BOD enter the pond?

13.227 The flow to a waste treatment pond is 180 gpm. If the BOD concentration of the water is 223 mg/L, how many pounds of BOD are applied to the pond daily?

13.228 The BOD concentration of the influent wastewater to a waste treatment pond is 200 mg/L. If the flow to the pond is 130 gpm, how many pounds of BOD are applied to the pond daily?

13.229 A 7.8 ac pond receives a flow of 219,000 gpd. If the influent flow has a BOD content of 192 mg/L, what is the organic loading rate in pounds per day per acre on the pond?

13.230 A pond has an average width of 420 ft and an average length of 740 ft. The flow to the pond is 167,000 gpd, with a BOD content of 145 mg/L. What is the organic loading in pounds per day per acre on the pond?

13.231 The flow to a pond is 72,000 gpd, with a BOD content of 128 mg/L. The pond has an average width of 240 ft and an average length of 391 ft. What is the organic loading rate in pounds per day per acre on the pond?

13.232 The maximum desired organic loading rate for a 15 ac pond is 22 lb BOD/day/ac. If the influent flow to the pond has a BOD concentration of 189 mg/L, what is the maximum desirable million gallons per day flow to the pond?

13.233 The BOD entering a waste treatment pond is 210 mg/L. If the BOD in the pond effluent is 41 mg/L, what is the BOD removal efficiency of the pool?

13.234 The influent of a waste treatment pond has a BOD content 267 mg/L. If the BOD content of the pond effluent is 140 mg/L, what is the BOD removal efficiency of the pond?

13.235 The BOD entering a waste treatment pond is 290 mg/L. If the BOD in the pond effluent is 44 mg/L, what is the BOD removal efficiency of the pond?

13.236 The BOD entering a waste treatment pond is 142 mg/L. If the BOD in the pond effluent is 58 mg/L, what is the BOD removal efficiency of the pond?

13.237 A 22 ac pond receives a flow of 3.6 ac-ft/day. What is the hydraulic loading rate on the pond in inches per day?

13.238 A 16 ac pond receives a flow of 6 ac-ft/day. What is the hydraulic loading rate on the pond in inches per day?

13.239 A waste treatment pond receives a flow of 2,410,000 gpd. If the surface area of the pond is 17 ac, what is the hydraulic loading in inches per day?

13.240 A waste treatment pond receives a flow of 1,880,000 gpd. If the surface area of the pond is 16 ac, what is the hydraulic loading in inches per day?

13.241 A 5 ac wastewater pond serves a population of 1,340 people. What is the population loading (people per acre) on the pond?

13.242 A wastewater pond serves a population of 5,580 people. If the pond covers 19.1 ac, what is the population loading (people per acre) on the pond?

13.243 A 0.8 MGD wastewater flow has a BOD concentration of 1,640 mg/L. Using an average of 0.3 lb BOD/day/person, what is the population equivalent of this wastewater flow?

13.244 A 257,000 gpd wastewater flow has a BOD content of 2,260 mg/L. Using an average of 0.2 lb BOD/day/person, what is the population equivalent of this flow?

13.245 A waste treatment pond has a total volume of 19 ac-ft. If the flow to the pond is 0.44 ac-ft/day, what is the detention time of the pond (in days)?

13.246 A waste treatment pond is operated at a depth of 8 ft. The average width of the pond is 450 ft, and the average length is 700 ft. If the flow to the pond is 0.3 MGD, what is the detention time in days?

13.247 The average width of the pond is 250 ft, and the average length is 400 ft. A waste treatment pond is operated at a depth of 6 ft. If the flow to the pond is 72,000 gpd, what is the detention time in days?

13.248 A waste treatment pond has an average length of 700 ft, an average width of 410 ft, and a water depth of 5 ft. If the flow to the pond is 0.48 ac-ft/day, what is the detention time for the pond?

13.249 A wastewater treatment pond has an average length of 720 ft, with an average width of 460 ft. If the flow rate to the pond is 310,000 gal each day and is operated at a depth of 6 ft, what is the hydraulic detention time in days?

13.250 What is the detention time for a pond receiving an influent flow rate of 0.50 ac-ft each day? The pond has an average length of 705 ft and an average width of 430 ft. The operating depth of the pond is 50 in.

13.251 A waste treatment pond has an average width of 395 ft and an average length of 698 ft. The influent flow rate to the pond is 0.16 MGD, with a BOD concentration of 171 mg/L. What is the organic loading rate to the pond in pounds per day per acre?

13.252 A pond 750 ft long and 435 ft wide receives an influent flow rate of 0.79 ac-ft/day. What is the hydraulic loading rate on the pond in inches per day?

13.253 The BOD concentration of the wastewater entering a pond is 192 mg/L. If the flow to the pond is 370,000 gpd, how many pounds per day BOD enter the pond?

13.254 A 9.1 ac pond receives a flow of 285,000 gpd. If the influent flow has a BOD content of 240 mg/L, what is the organic rate in pounds per day per acre on the pond?

13.255 The BOD entering a waste treatment pond is 220 mg/L. If the BOD concentration in the pond effluent is 44 mg/L, what is the BOD removal efficiency of the pond?

13.256 A 22 ac pond receives a flow of 3.8 ac-ft/day. What is the hydraulic loading on the pond in inches per day?

13.257 The BOD entering a waste treatment pond is 166 mg/L. If the BOD concentration in the pond effluent is 73 mg/L, what is the BOD removal efficiency of the pond?

13.258 The flow to a waste treatment pond is 210 gpm. If the BOD concentration of the water is 222 mg/L, how many pounds of BOD are applied to the pond daily?

13.259 The flow to a pond is 80,000 gpd, with a BOD content of 135 mg/L. The pond has an average width of 220 ft and an average length of 400 ft. What is the organic loading rate in pounds per day per acre on the pond?

13.260 A waste treatment pond receives a flow of 1,980,000 gpd. If the surface area of the pond is 21 ac, what is the hydraulic loading in inches per day?

13.261 A wastewater pond serves a population of 6,200 people. If the area of the pond is 22 ac, what is the population loading on the pond?

13.262 A waste treatment pond has a total volume of 18.4 ac-ft. If the flow to the pond is 0.52 ac-ft/day, what is the detention time of the pond in days?

13.263 A 0.9 MGD wastewater flow has a BOD concentration of 2,910 mg/L. Using an average of 0.4 lb/day BOD/person, what is the population equivalent of this wastewater flow?

13.264 A waste treatment pond is operated at a depth of 6 ft. The average width of the pond is 440 ft, and the average length is 730 ft. If the flow to the pond is 0.45 MGD, what is the detention time in days?

13.265 Determine the chlorinator setting (pounds per day) needed to treat a flow of 4.6 MGD with a chlorine dose of 3.4 mg/L.

13.266 The desired dosage for a dry polymer is 11 mg/L. If the flow to be treated is 1,680,000 gpd, how many pounds per day polymer will be required?

13.267 To neutralize a sour digester, 1 lb of lime is to be added for every pound of volatile acids in the digester sludge. If the digester contains 200,000 gal of sludge, with a volatile acid level of 2,200 mg/L, how many pounds of lime should be added?

13.268 A total of 320 lb of chlorine was used during a 24 hr period to chlorinate a flow of 5.12 MGD. At this pounds per day dosage rate, what was the milligrams per liter dosage rate?

13.269 The secondary effluent is tested and found to have a chlorine demand of 4.9 mg/L. If the desired residual is 0.8 mg/L, what is the desired chlorine dose in milligrams per liter?

13.270 The chlorine dosage for a secondary effluent is 8.8 mg/L. If the chlorine residual after 30 min contact time is found to be 0.9 mg/L, what is the chlorine demand expressed in milligrams per liter?

13.271 The chlorine demand of a secondary effluent is 7.9 mg/L. If a chlorine residual of 0.6 mg/L is desired, what is the desired chlorine dosage in milligrams per liter?

13.272 What should the chlorinator setting be (pounds per day) to treat a flow of 4.0 MGD if the chlorine demand is 9 mg/L and a chlorine residual of 1.7 mg/L is desired?

13.273 A total chlorine dosage of 11.1 mg/L is required to treat a water unit. If the flow is 2.88 MGD and the hypochlorite has 65% available chlorine, how many pounds per day of hypochlorite will be required?

13.274 The desired dose of polymer is 9.8 mg/L. The polymer provides 60% active polymer. If a flow of 4.1 MGD is to be treated, how many pounds per day of the polymer compound will be required?

13.275 A wastewater flow of 1,724,000 gpd requires a chlorine dose of 19 mg/L. If hypochlorite (65% available chlorine) is to be used, how many pounds per day of hypochlorite are required?

13.276 A total of 950 lb of 65% hypochlorite is used in a day. If the flow rate is treated to 5.65 MGD, what is the chlorine dosage in milligrams per liter?

13.277 If a total of 12 oz of dry polymer is added to 16 gal of water, what is the percent strength (by weight) of the polymer?

13.278 How many pounds of dry polymer must be added to 24 gal of water to make a 0.9% polymer solution?

13.279 If 160 g of dry polymer are dissolved in 12 gal of water, what percent strength is the solution? (1 g = 0.0022 lb.)

13.280 A 10% liquid polymer is to be used in making up a polymer solution. How many pounds of liquid polymer should be mixed with water to produce 172 lb of a 0.5% polymer solution?

13.281 A 10% liquid polymer will be used in making up a solution. How many gallons of liquid polymer should be added to the water to make up 55 gal of a 0.3% polymer solution? The liquid polymer has a specific gravity of 1.25, and assume the polymer solution has a specific gravity of 1.0.

13.282 If 26 lb of a 10% solution are mixed with 110 lb of a 0.5% strength solution, what is the percent strength of the solution mixture?

13.283 If 6 gal of a 12% strength solution are added to 30 gal of a 0.4% strength solution, what is the percent strength of the solution mixture? Assume the 12% solution weighs 10.2 lb/gal and the 0.3% strength solution weighs 8.4 lb/gal.

13.284 If 12 gal of a 10% strength solution are mixed with 42 gal of a 0.28% strength solution, what is the percent strength of the solution mixture? Assume the 10% solution weighs 10.2 lb/gal and the 0.26% solution weighs 8.34 lb/gal.

13.285 Jar tests indicate that the best liquid alum dose for a water unit is 10 mg/L. The flow to be treated is 4.10 MGD. Determine the gallons per day setting for the liquid alum chemical feeder if the liquid alum contains 5.88 lb of alum per gallon of solution.

13.286 Jar tests indicate that the best liquid alum dose for a water unit is 8 mg/L. The flow to be treated is 1,440,000 gpd. Determine the gallons per day setting for the liquid alum chemical feeder if the liquid alum contains 6.15 lb of alum per gallon of solution.

13.287 Jar tests indicate that the best liquid alum dose for a water unit is 11 mg/L. The flow to be treated is 2.13 MGD. Determine the gallons per day setting for the liquid alum chemical feeder if the liquid alum is a 60% solution. Assume the alum solution weighs 8.34 lb/gal.

13.288 The flow to the plant is 4,440,000 gpd. Jar testing indicates that the optimum alum dose is 9 mg/L. What should the gallons per day setting be for the solution feeder if the alum solution is a 60% solution? Assume the solution weighs 8.34 lb/gal.

13.289 The required chemical pumping rate has been calculated to be 30 gpm. If the maximum pumping rate is 80 gpm, what should the percent stroke setting be?

13.290 The required chemical pumping rate has been calculated to be 22 gpm. If the maximum pumping rate is 80 gpm, what should the percent stroke setting be?

13.291 The required chemical pumping rate has been determined to be 14 gpm. What is the percent stroke setting if the maximum rate is 70 gpm?

13.292 The maximum pumping rate is 110 gpm. If the required pumping rate is 40 gpm, what is the percent stroke setting?

13.293 The desired solution feed rate was calculated to be 35 gpd. What is this feed rate expressed as milliliters per minute?

13.294 The desired solution feed rate was calculated to be 45 gpd. What is this feed rate expressed as milliliters per minute?

13.295 The optimum polymer dose has been determined to be 9 mg/L. The flow to be treated is 910,000 gpd. If the solution to be used contains 60% active polymer, what should the solution chemical feeder setting be in milliliters per minute? Assume the polymer solution weighs 8.34 lb/gal.

13.296 The flow to be treated is 1,420,000 gpd. The optimum polymer dose has been determined to be 11 mg/L. If the solution to be used contains 60% active polymer, what should the solution chemical feeder setting be in milliliters per minute. Assume the polymer solution weighs 8.34 lb/gal.

13.297 Calculate the actual chemical feed rate, in pounds per day, if a container is placed under a chemical feeder and a total of 2.1 lb is collected during a 30 min period.

13.298 Calculate the actual chemical feed rate, in pounds per day, if a bucket is placed under a chemical feeder and total of 1 lb, 8 oz is collected during a 30 min period.

13.299 To calibrate a chemical feeder, a container is first weighed (12 oz), then placed under the chemical feeder. After 30 min, the container is weighed again. If the weight of the container with chemical is 2.10 lb, what is the actual chemical feed rate in pounds per day?

13.300 A chemical feeder is to be calibrated. The container to be used to collect chemical is placed under the chemical feeder and weighed (0.5 lb). After 30 min, the weight of the container and chemical is found to be 2.5 lb. Based on this test, what is the actual chemical feed rate in pounds per day?

13.301 A calibration test is conducted for a solution chemical feeder. For 5 min, the solution feeder delivers a total of 770 mg/L. The polymer solution is 1.4% solution. What is the pounds per day polymer feed rate? Assume the polymer solution weighs 8.34 lb/gal.

13.302 A calibration test is conducted for a solution chemical feeder. During the 5 min test, the pump delivered 900 mL of the 1.2% polymer solution. What is the polymer dosage rate in pounds per day? Assume the polymer solution weighs 8.34 lb/gal.

13.303 A calibration test is conducted for a solution chemical feeder. During a 5 min test, the pump delivered 610 mL of 1.3% polymer solution. The specific gravity of the polymer solution is 1.2. What is the polymer dosage rate in pounds per day?

13.304 During a 5 min test, the pump delivered 800 mL of a 0.5% polymer solution. A calibration test is conducted for the solution chemical feeder. The specific gravity of the polymer solution is 1.15. What is the polymer dosage rate in pounds per day?

13.305 A pumping rate calibration test is conducted for a 3 min period. The liquid level in the 4 ft diameter solution tank is measured before and after the test. If the level drops 1.5 ft during a 3 min test, what is the pumping rate in gallons per minute?

13.306 A pumping rate calibration test is conducted for a 5 min period. The liquid level in the 5 ft diameter tank is measured before and after the test. If the level drops 15 in during the test, what is the pumping rate in gallons per minute?

13.307 A pump test indicates that a pump delivers 30 gpm during a 4 min pumping test. The diameter of the solution tank is 5 ft. What was the expected drop, in feet, in solution level during the pumping test?

13.308 The liquid level in the 5 ft diameter solution tank is measured before and after the test. A pumping rate calibration test is conducted for a 3 min period. If the level drops 18 in during the test, what is the pumping rate in gallons per minute?

13.309 The amount of chemical used for each day during a week is given in the following. Based on this data, what was the average pounds per day chemical use during the week?

Monday: 81 lb/day
Tuesday: 73 lb/day
Wednesday: 74 lb/day
Thursday: 66 lb/day
Friday: 79 lb/day
Saturday: 83 lb/day
Sunday: 81 lb/day

13.310 The average chemical use at a plant is 115 lb/day. If the chemical inventory in stock is 2,300 lb, how many days' supply is this?

13.311 The chemical inventory in stock is 1,002 lb. If the average chemical use at a plant is 66 lb/day, how many days' supply is this?

13.312 The average gallons of polymer solution used each day at a treatment plant is 97 gpd. A chemical feed tank has a diameter of 5 ft and contains solution to a depth of 3 ft, 4 in. How many days' supply is represented by the solution in the tank?

13.313 The desired dosage for a dry polymer is 11 mg/L. If the flow to be treated is 3,250,000 gpd, how many pounds per day polymer will be required?

13.314 A total chlorine dosage of 7.1 mg/L is required to treat a particular water unit. If the flow is 3.24 MGD and the hypochlorite has 65% available chlorine, how many pounds per day of hypochlorite will be required?

13.315 How many pounds of dry polymer must be added to 32 gal of water to make a 0.2% polymer solution?

13.316 Calculate the actual chemical feed rate, in pounds per day, if a bucket is placed under a chemical feeder and a total of 1.9 lb is collected during a 30 min period.

13.317 Jar tests indicate that the best liquid alum dose for a water unit is 12 mg/L. The flow to be treated is 2,750,000 gpd. Determine the gallons per day setting for the liquid alum chemical feeder if the liquid alum contains 5.88 lb of alum per gallon of solution.

13.318 A total of 379 lb of chlorine was used during a 24 hr period to chlorinate a flow of 5,115,000 gpd. At this pounds per day dosage rate, what was the milligrams per liter dosage rate?

13.319 To calibrate a chemical feeder, a container is first weighed (12 oz) then placed under the chemical feeder. After 30 min, the bucket is weighed again. If the weight of the bucket with the chemical is 2 lb, 6 oz, what is the actual chemical feed rate in pounds per day?

13.320 The flow to a plant is 3,244,000 gpd. Jar testing indicates that the optimum alum dose is 10 mg/L. What should the gallons per day settling be for the solution feeder if the alum solution is a 60% solution? Assume the alum solution weighs 8.34 lb/gal.

13.321 The desired chemical pumping rate has been calculated at 32 gpm. If the minimum pumping rate is 90 gpm, what should the percent stroke setting be?

13.322 The chlorine dosage for a secondary effluent is 7.8 mg/L. If the chlorine residual after 30 min contact time is found to be 0.5 mg/L, what is the chlorine demand expressed in milligrams per liter?

13.323 How many gallons of 12% liquid polymer should be mixed with water to produce 60 gal of a 0.4% polymer solution? The density of the polymer liquid is 9.6 lb/gal. Assume the density of the polymer solution is 8.34 lb/gal.

13.324 A calibration test is conducted for a solution chemical feeder. For 5 min, the solution feeder delivers a total of 660 mL. The polymer solution is a 1.2% solution. What is the pounds per day feed rate? Assume the polymer solution weighs 8.34 lb/gal.

13.325 A pump operates at a rate of 30 gpm. How many feet will the liquid level be expected to drop after a 5 min pumping test if the diameter of the solution tank is 6 ft?

13.326 The desired chemical pumping rate has been calculated to be 20 gpm. If the maximum pumping rate is 90 gpm, what should the percent stroke settling be?

13.327 What should the chlorinator setting be (in pounds per day) to treat a flow of 4.3 MGD if the chlorine demand is 8.7 mg/L and a chlorine residual of 0.9 mg/L is desired?

13.328 The average chemical use at a plant is 90 lb/day. If the chemical inventory in stock is 2,100 lb, how many days' supply is this?

13.329 The desired solution feed rate was calculated to be 50 gpd. What is this feed rate expressed as milliliters per minute?

13.330 A calibration test is conducted for a solution chemical feeder. During a 5 min test, the pump delivered 888 mL of the 0.9% polymer solution. What is the polymer dosage rate in pounds per day? Assume the polymer solution weighs 8.34 lb/gal.

13.331 The flow to be treated is 3,220,000 gpd. The optimum polymer dose has been determined be 9 mg/L. If the solution to be used contains 60% active polymer, what should the solution chemical feeder setting be in milliliters per minute?

13.332 A pumping calibration test is conducted for a 3 min period. The liquid level in the 4 ft diameter solution tank is measured before and after the test. If the level drops 15 in during the 3 min test, what is the pumping rate in gallons per minute?

13.333 A wastewater flow of 3,115,000 gpd requires a chlorine dose of 11.1 mg/L. If hypochlorite (65% available chlorine) is to be used, how many pounds per day of hypochlorite are required?

13.334 If 6 gal of a 12% strength solution are mixed with 22 gal of a 0.3% strength solution, what is the percent strength of the solution mixture? Assume the 12% solution weighs 11.2 lb/gal and the 0.3% solution weighs 8.34 lb/gal.

13.335 A primary clarifier receives a flow of 4.82 MGD, with a suspended solids concentration of 291 mg/L. If the clarifier effluent has a suspended solids concentration of 131 mg/L, how many pounds of dry solids are generated daily?

13.336 The suspended solids concentration of the primary influent is 315 mg/L, and that of the primary effluent is 131 mg/L. How many pounds of dry solids are produced if the flow is 3.9 MGD?

13.337 The 2.1 MGD influent to the secondary system has a BOD concentration of 260 mg/L. The secondary effluent contains 125 mg/L BOD. If the bacterial growth rate, Y-value, for this plant is 0.6 lb SS/lb BOD removed, how many pounds of dry solids are produced each day by the secondary system?

13.338 The Y-value for a treatment plant secondary system is 0.66 lb SS/lb BOD removed. The influent to the secondary system is 2.84 MGD. If the BOD concentration of the secondary influent is 288 mg/L and the effluent BOD is 131 mg/L, how many pounds of dry solids are produced each day by the secondary system?

13.339 The total weight of a sludge sample is 31 g (sludge sample only, not the dish). If the weight of the solids after drying is 0.71 g, what is the percent total solids of the sludge?

13.340 A total of 8,520 lb/day SS is removed from a primary clarifier and pumped to a sludge thickener. If the sludge has a solids content of 4.2%, how many pounds per day sludge are pumped to the thickener?

13.341 A total of 9,350 gal of sludge is pumped to a digester. If the sludge has a 5.5% solids content, how many pounds per day solids are pumped to the digester? Assume the sludge weighs 8.34 lb/gal.

13.342 It is anticipated that 1,490 lb/day SS will be pumped from the primary clarifier of a new plant. If the primary clarifier sludge has a solids content of 5.3%, how man gallons per day sludge will be pumped from the clarifier? Assume the sludge weighs 8.34 lb/gal.

13.343 A primary sludge has a solids content of 4.4%. If 900 lb/day suspended solids are pumped from the primary clarifier, how many gallons per day sludge will be pumped from the clarifier? Assume the sludge weighs 8.34 lb/gal/

13.344 A total of 20,100 lb/day sludge is pumped to a thickener. The sludge has 4.1% solids content. If the sludge is concentrated to 6% solids, what will be the expected pounds per day sludge flow from the thickener?

13.345 A primary clarifier sludge has 5.1% solids content. If 2,910 gpd of primary sludge are pumped to a thickener and the thickened sludge has a solids content of 6%, what would be the expected gallons per day flow of the thickened sludge? Assume the primary sludge weighs 8.34 lb/gal and the thickened sludge weighs 8.64 lb/gal.

13.346 A primary clarifier sludge has 3.4% solids content. A total of 12,400 lb/day sludge is pumped to a thickener. If the sludge has been concentrated to 5.4% solids, what will be the pounds per day sludge flow from the thickener?

13.347 The sludge from a primary clarifier has a solids content of 4.1%. The primary sludge is pumped at a rate of 6,100 gpd to a thickener. If the thickened sludge has a solids content of 6.4%, what is the anticipated gallons per day sludge flow through the thickener? Assume the primary sludge weighs 8.34 lb/gal and the secondary sludge weighs 8.6 lb/gal.

13.348 A gravity thickener 28 ft in diameter receives a flow of 70 gpm primary sludge combined with an 82 gpm secondary effluent flow. What is the hydraulic loading on the thickener in gallons per day per square foot?

13.349 The primary sludge flow to a gravity thickener is 90 gpm. This is blended with a 72 gpm secondary effluent flow. If the thickener has a diameter of 28 ft, what is the hydraulic loading rate in gallons per day per square foot?

13.350 A primary sludge flow equivalent to 122,000 gpd is pumped to a 44 ft diameter gravity thickener. If the solids concentration of the sludge is 4.1%, what is the solids loading in pounds per day per square foot?

13.351 What is the solids loading on a gravity thickener (in pounds per day per square foot) if the primary sludge flow to the 32 ft diameter gravity thickener is 60 gpm, with a solids concentration of 3.8%?

13.352 A gravity thickener 46 ft in diameter has a sludge blanket depth of 3.8 ft. If sludge is pumped from the bottom of the thickener at the rate of 28 gpm, what is the sludge detention time (in days) in the thickener?

13.353 A gravity thickener 40 ft in diameter has a sludge blanket depth of 4.3 ft. If the sludge is pumped from the bottom of the thickener at a rate of 31 gpm, what is the sludge detention time, in hours, in the thickener?

13.354 What is the efficiency of the gravity thickener if the influent flow to the thickener has a sludge solids concentration of 4% and the effluent flow has a sludge solids concentration of 0.9%?

13.355 The sludge solids concentration of the influent flow to a gravity thickener is 3.3%. If the sludge withdrawn from the bottom of the thickener has a sludge solids concentration of 8.4%, what is the concentration factor?

13.356 The influent flow to a gravity thickener has a sludge solids concentration of 3.1%. What is the concentration factor if the sludge solids concentration of the sludge withdrawn from the thickener is 8.0%?

13.357 Given the following data, determine whether the sludge blanket in the gravity thickener is expected to increase, decrease, or remain the same.

Sludge pumped to thickener: 130 gpm
Thickener sludge pumped from thickener: 50 gpm
Primary sludge solids: 3.6%
Thickened sludge solids: 8.1%
Thickener effluent suspended solids: 590 mg/L

13.358 Given the data that follows, (a) determine whether the sludge blanket in the gravity thickener is expected to increase, decrease, or remain the same, and (b) if there is an increase or decrease, how many pounds per day should this change be?

Sludge pumped to thickener: 110 gpm
Thickener sludge pumped from thickener: 65 gpm
Primary sludge solids: 3.6%
Thickened sludge solids: 7.1%
Thickener effluent suspended solids: 520 mg/L

13.359 If solids are being stored at a rate of 9,400 lb/day in a 30 ft diameter gravity thickener, how many hours will it take the sludge blanket to rise 1.8 ft? The solids concentration of the thickened sludge is 6.6%.

13.360 Solids are being stored at a rate of 14,000 lb/day in a 30 ft diameter gravity thickener. How many hours will it take the sludge blanket to rise 2.5 ft? The solids concentration of the thickened sludge is 8%.

13.361 After several hours of startup of a gravity thickener, the sludge blanket level is measured at 2.6 ft. The desired sludge blanket level is 6 ft. If the sludge solids are entering the thickener at a rate of 60 lb/min, what is the desired sludge withdrawal rate in gallons per minute? The thickened sludge solids concentration is 5.6%.

13.362 The sludge blanket level is measured at 3.3 ft after several hours of startup of a gravity thickener. If the desired sludge blanket level is 7 ft and the sludge solids are entering the thickener at a rate of 61 lb/min, what is the desired sludge withdrawal rate in gallons per minute? The thickened sludge solids concentration is 5.6%.

13.363 A dissolved air flotation (DAF) thickener receives a sludge flow of 910 gpm. If the DAF unit is 40 ft in diameter, what is the hydraulic loading rate in gallons per minute per square foot?

13.364 A dissolved air floatation (DAF) thickener 32 ft in diameter receives a sludge flow of 660 gpm. What is the hydraulic loading rate in gallons per minute per square foot?

13.365 The sludge flow to a 40 ft diameter dissolved air thickener is 170,000 gpd. If the influent waste activated sludge has a suspended solids concentration of 8,420, what is the solids loading rate in pounds per hour per square foot? Assume the sludge weighs 8.34 lb/gal.

13.366 The sludge flow to a dissolved air flotation thickener is 120 gpm, with a suspended solids concentration of 0.7%. If the DAF unit is 65 ft long and 20 ft wide, what is the solids loading rate in pounds per hour per square foot? Assume the sludge weighs 8.34 lb/gal.

13.367 The air rotameter indicates 9 cfm is supplied to the dissolved air flotation thickener. What is this air supply expressed as pounds per hour?

13.368 The air rotameter for the dissolved air flotation thickener indicates 12 cfm is supplied to the DAF unit. What is this air supply expressed as pounds per hour?

13.369 A dissolved air flotation thickener receives an 85 gpm flow of waste activated sludge with a solids concentration of 8,600 mg/L. If air is supplied at a rate of 8 cfm, what is the air-to-solids ratio?

13.370 The sludge flow to a dissolved air flotation thickener is 60 gpm, with a suspended solids concentration of 7,800 mg/L. If the air supplied to the DAF unit is 5 cfm, what is the air-to-solids ratio?

13.371 A dissolved air flotation thickener receives a sludge flow of 85 gpm. If the recycle rate is 90 gpm, what is the percent recycle rate?

13.372 The desired percent recycle rate for a dissolved air flotation unit is 112%. If the sludge flow to the thickener is 70 gpm, what should the recycle flow be in million gallons per day?

13.373 An 80 ft diameter DAF thickener receives a sludge flow, with a solids concentration of 7,700 mg/L. If the effluent solids concentration is 240 mg/L, what is the solids removal efficiency?

13.374 The solids concentration of the influent sludge to a dissolved air flotation unit is 8,410 mg/L. If the thickened sludge solids concentration is 4.8%, what is the concentration factor?

13.375 A disc centrifuge receives a waste activated sludge flow of 40 gpm. What is the hydraulic loading on the unit in gallons per hour?

13.376 The waste activated sludge flow to a scroll centrifuge thickener is 86,400 gpd. What is the hydraulic loading on the thickener in gallons per hour?

13.377 The waste activated sludge flow to a basket centrifuge is 70 gpm. The basket run time is 30 min until the basket is full of solids. If it takes 1 min to skim the solids out of the unit, what is the hydraulic loading rate on the unit in gallons per hour?

13.378 The sludge flow to a basket centrifuge is 78,000 gpd. The basket run time is 25 min until the flow to the unit must be stopped for the skimming operation. If skimming takes 2 min, what is the hydraulic loading on the unit in gallons per hour?

13.379 A scroll centrifuge receives a waste activated sludge flow of 110,000 gpd, with a solids concentration of 7,600 mg/L. What is the solids loading in pounds per hour to the centrifuge?

13.380 The sludge flow to a basket thickener is 80 gpm, with a solids concentration of 7,500 mg/L. The basket operates 30 min before the flow must be stopped to the unit during the 2 min skimming operation. What is the solids loading in pounds per hour to the centrifuge?

13.381 A basket centrifuge with a 32 cu ft capacity receives a flow of 70 gpm. The influent sludge solids concentration is 7,400 mg/L. The average solids concentration within the basket is 6.6%. What is the feed time (minutes) for the centrifuge?

13.382 A basket centrifuge thickener has a capacity of 22 cu ft. The 55 gpm sludge flow to the thickener has a solids concentration of 7,600 mg/L. The average solids concentration within the basket is 9%. What is the feed time (minutes) for the centrifuge?

13.383 The influent sludge solids concentration to a disc centrifuge is 8,000 mg/L. If the sludge solids concentration of the centrifuge effluent (centrate) is 800 mg/L, what is the sludge solids removal efficiency?

13.384 A total of 16 cu ft of skimmed sludge and 4 cu ft of knifed sludge is removed from a basket centrifuge. If the skimmed sludge has a solids concentration of 4.4% and the knifed sludge has a solids concentration of 8.0%, what is the percent solids concentration of the sludge mixture?

13.385 A total of 12 cu ft of skimmed sludge and 4 cu ft of knifed sludge is removed from a basket centrifuge. If the skimmed sludge has a solids concentration of 3.8% and the knifed sludge has a solids concentration of 8.0%, what is the percent solids concentration of the sludge mixture?

13.386 A solid bowl centrifuge receives 48,400 gal of sludge daily. The sludge concentration before thickening is 0.8. How many pounds of solids are received each day?

13.387 A gravity thickener receives a primary sludge flow rate of 170 gpm. If the thickener has a diameter of 24 ft, what is the hydraulic loading rate in gallons per day per square foot?

13.388 The primary sludge flow rate to a 40 ft diameter gravity thickener is 240 gpm. If the solids concentration is 1.3%, what is the solids loading rate in pounds per square foot?

13.389 Waste activated sludge is pumped to a 34 ft diameter dissolved air flotation thickener at a rate of 690 gpm. What is the hydraulic loading rate in gallons per minute per square foot?

13.390 Waste activated sludge is pumped to a 30 ft diameter dissolved air flotation thickener at a rate of 130 gpm. If the concentration of solids is 0.98%, what is the solids loading rate in pounds per hour per square foot?

13.391 The suspended solids content of the primary influent is 305 mg/L and that of the primary effluent is 124 mg/L. How many pounds of solids are produced during a day that the flow is 3.5 MGD?

13.392 The total weight of a sludge sample is 32 g (sample weight only, not including the weight of the dish). If the weight of the solids after drying is 0.66 g, what is the percent total solids of the sludge?

13.393 A primary clarifier sludge has 3.9% solids content. If 3,750 gpd primary sludge is pumped to a thickener and the thickened sludge has a solids content of 8%, what would be the expected gallons per day flow of thickened sludge? Assume both sludges weigh 8.34 lb/gal.

13.394 A total of 9,550 gal of sludge is pumped to a digester daily. If the sludge has a 4.9% solids content, how many pounds per day solids are pumped to the digester? Assume the sludge weighs 8.34 lb/gal.

13.395 The 2.96 MGD influent to a secondary system has a BOD concentration of 170 mg/L. The secondary effluent contains 38 mg/L BOD. If the bacteria growth rate, Y-value, for this plant is 0.5 lb SS/lb BOD removed, what is the estimated pounds of dry solids produced each day by the secondary system?

13.396 A gravity thickener 42 ft in diameter has a sludge blanket depth of 5 ft. If sludge is pumped from the bottom of the thickener at the rate of 32 gpm, what is the sludge detention time (in hours) in the thickener?

13.397 The solids concentration of the influent flow to a gravity thickener is 3.1%. If the sludge withdrawn from the bottom of the thickener has a solids concentration of 7.7%, what is the concentration factor?

13.398 A 75 ft diameter DAF thickener receives a sludge flow with a solids concentration of 7,010 mg/L. If the effluent solids concentration is 230 mg/L, what is the solids removal efficiency?

13.399 The primary sludge flow to a gravity thickener is 70 gpm. This is blended with a 100 gpm secondary effluent flow. If the thickener has a diameter of 30 ft, what is the hydraulic loading rate in gallons per day per square foot?

13.400 What is the efficiency of a gravity thickener if the influent flow to the thickener has a sludge solids concentration of 3.3% and the effluent flow has a sludge solids concentration of 0.3%.

13.401 The air supplied to a dissolved air flotation thickener is 9 cfm. What is this air supply expressed as pounds per hour?

13.402 Given the following data, determine whether the sludge blanket in the gravity thickener will increase, decrease, or remain the same.

Sludge pumped to thickener: 110 gpm
Thickened sludge pumped from thickener: 50 gpm
Primary sludge solids: 4%
Thickened sludge solids: 7.7%
Thickener effluent suspended solids: 700 mg/L

13.403 What is the solids loading on a gravity thickener, in pounds per day per square foot, if the primary sludge flow to the 32 ft diameter thickener is 60 gpm, with a solids concentration of 4.1%.

13.404 A dissolved air flotation unit is 60 ft long and 14 ft wide. If the unit receives a sludge flow of 190,000 gpd, what is the hydraulic loading rate in gallons per minute per square foot?

13.405 If solids are being stored at a rate of 9,400 lb/day in a 26 ft diameter gravity thickener, how many hours will it take the sludge blanket to rise to 2.6 ft? The solids concentration of the thickened sludge is 6.9%.

13.406 The waste activated sludge flow to a scroll centrifuge thickener is 84,000 gpd. What is the hydraulic loading rate on the thickener in gallons per hour?

13.407 The sludge flow to a DAF thickener is 110 gpm. The solids concentration of the sludge is 0.81%. If the air supplied to the DAF unit is 6 cfm, what is the air-to-solids ratio?

13.408 The desired percent recycle for a DAF unit is 112%. If the sludge flow to the thickener is 74 gpm, what should the recycle flow be in million gallons per day?

13.409 The sludge flow to a basket centrifuge is 79,000 gpd. The basket run time is 32 min until the flow to the unit must be stopped for the skimming operation. If skimming takes 2 min, what is the hydraulic loading on the unit in gallons per hour?

13.410 After several hours of startup of a gravity thickener, the sludge blanket level is measured at 2.5 ft. The desired sludge blanket level is 6 ft. If the sludge solids are entering the thickener at a rate of 48 lb/min, what is the desired sludge withdrawal rate in gallons per minute? The thickened sludge solids concentration is 8%.

13.411 A scroll centrifuge receives a waste activated sludge flow of 110,000 gpd, with a solids concentration of 7,110 mg/L. What is the solids loading to the centrifuge in pounds per hour?

13.412 A basket centrifuge with a 34 cu ft capacity receives a flow of 70 gpm. The influent sludge solids concentration is 7,300 mg/L. The average solid concentration within the basket is 6.6%. What is the feed time for the centrifuge in minutes?

13.413 The sludge flow to a basket thickener is 100 gpm, with a solids concentration of 7,900 mg/L. The basket operates 24 min before the flow must be stopped to the unit during the 1.5 min skimming operation. What is the solids loading to the centrifuge in pounds per hour?

13.414 A total of 12 cu ft of skimmed sludge and 5 cu ft of knifed sludge is removed from a basket centrifuge. If the skimmed sludge has a solids concentration of 3.9% and the knifed sludge has a solids concentration of 7.8%, what is the solids concentration of the sludge mixture?

13.415 A primary sludge flow of 4,240 gpd with a solids content of 5.9% is mixed with a thickened secondary sludge flow of 6,810 gpd, with a solids content of 3.5%. What is the percent solids content of the mixed sludge flow? Assume both sludges weigh 8.34 lb/gal.

13.416 Primary and thickened secondary sludges are to be mixed and sent to the digester. The 3510 gpd primary sludge has a solids content of 5.2%, and the 5,210 gpd thickened secondary sludge has a solids content of 4.1%. What would be the percent solids content of the mixed sludge? Assume both sludges weigh 8.3 lb/gal.

13.417 A primary sludge flow of 3,910 gpd with a solids content of 6.3% is mixed with a thickened secondary sludge flow of 6,690 gpd, with a solids content of 4.9%. What is the percent solids of the combined sludge flow? Assume both sludges weigh 8.34 lb/gal.

13.418 Primary and secondary sludges are to be mixed and sent to the digester. The 2,510 gpd primary sludge has a solids content of 4.3%, and the 3,600 gpd thickened secondary sludge has a solids content of 6.1%. What would be the percent solids content of the mixed sludge? Assume the 4.3% sludge weighs 8.35 lb/gal and the 6.1% sludge weighs 8.60 lb/gal.

13.419 A piston pump discharges a total of 0.9 gal per stroke. If the pump operates at 30 strokes per minute, what is the gallons per minute pumping rate? Assume the piston is 100% efficient and displaces 100% of its volume each stroke.

13.420 A sludge pump has a bore of 10 in and a stroke length of 3 in. If the pump operates at 30 strokes per minute, how many gallons per minute are pumped? Assume 100% efficiency.

13.421 A sludge pump has a bore of 8 in and a stroke setting of 3 in. The pump operates at 32 strokes per minute. If the pump operates a total of 120 min during a 24 hr period, what is the gallons per day pumping rate? Assume 100% efficiency.

13.422 A sludge pump has a bore of 12 in and a stroke setting of 4 in. The pump operates at 32 strokes per minute. If the pump operates a total of 140 min during a 24 hr period, what is the gallons per day pumping rate? Assume 100% efficiency.

13.423 The flow to a primary clarifier is 2.5 MGD. The influent suspended solids concentration is 240 mg/L, and the effluent suspended solids concentration is 110 mg/L. If the sludge to be removed from the clarifier has solids content of 3.5% and the sludge pumping rate is 32 gpm, how many minutes per hour should the pump operate?

13.424 The suspended solids concentration of the 1,870,000 gpd influent flow to a primary clarifier is 210 mg/L. The primary clarifier effluent flow suspended solids concentration is 90 mg/L. If the sludge to be removed from the clarifier has solids content of 3.6% and the sludge pumping rate is 28 gpm, how many minutes per hour should the pump operate?

13.425 A primary clarifier receives a flow of 3,480,000 gpd, with a suspended solids concentration of 220 mg/L. The clarifier effluent has a suspended solids concentration of 96 mg/L. If the sludge to be removed from the clarifier has solids content of 4.0% and the sludge pumping rate is 38 gpm, how many minutes per hour should the pump operate?

13.426 If 8,620 lb/day of solids with a volatile solids content of 66% are sent to the digester, how many pounds per day volatile solids are sent to the digester?

13.427 If 2,810 lb/day of solids with a volatile solids content of 67% are sent to the digester, how many pounds per day volatile solids are sent to the digester?

13.428 A total of 3,720 gpd of sludge is to be pumped to the digester. If the sludge has 5.8% solids content with 70% volatile solids, how many pounds per day volatile solids are pumped to the digester? Assume the sludge weighs 8.34 lb/gal.

13.429 The sludge has 7% solids content with 67% volatile solids. If a total of 5,115 gpd of sludge is to be pumped to the digester, how many pounds per day volatile solids are pumped in the digester?

13.430 A digester has a capacity of 295,200 gal. If the digester seed sludge is to be 24% of the digester capacity, how many gallons of seed sludge will be required?

13.431 A 40 ft diameter digester has a side water depth of 24 ft. If the seed sludge to be used is 21% of the tank capacity, how many gallons of seed sludge will be required?

13.432 A 50 ft diameter digester has a side water depth of 20 ft. If 62,200 gal of seed sludge are to be used in starting up the digester, what percent of the digester volume will be seed sludge?

13.433 A digester 40 ft in diameter has a side water depth of 18 ft. If the digester seed sludge is to be 20% of the digester capacity, how many gallons of seed sludge will be required?

13.434 A total of 66,310 lb/day of sludge is pumped to a 120,000 gal digester. The sludge being pumped to the digester has total solids content of 5.3% and a volatile solids content of 70%. The sludge in the digester has a solids content of 6.3%, with a 56% volatile solids content. What is the volatile solids loading on the digester in the volatile solids added per day per pound volatile solids in the digester?

13.435 A total of 22,310 gal of digested sludge is in a digester. The digested sludge contains 6.2% total solids and 55% volatile solids. To maintain volatile solids loading ratio of 0.06 lb volatile solids added per day per pound volatile solids under digestion, how many pounds volatile solids may enter the digester daily?

13.436 A total of 60,400 lb/day sludge is pumped to a 96,000 gal digester. The sludge pumped to the digester has a total solids content of 5.4% and a volatile solids content of 67%. The sludge in the digester has a solids content of 5%, with a 58% volatile solids content. What is the volatile solids loading on the digester in pounds volatile solids added per day per pound volatile solids in digester?

13.437 The raw sludge flow to the new digester is expected to be 900 gpd. The raw sludge contains 5.5% solids and 69% volatile solids. The desired volatile solids loading ratio is 0.07 lb volatile solids added per day per pound volatile solids in the digester. How many gallons of seed sludge will be required if the seed sludge contains 8.2% solids, with a 52% volatile solids content? Assume the seed sludge weighs 8.80 lb/gal.

13.438 A digester 50 ft in diameter, with a water depth of 22 ft, receives 86,100 lb/day raw sludge. If the sludge contains 5% solids with 70% volatile matter, what is the digester loading in pounds volatile solids added per day per cubic foot volume?

13.439 What is the digester loading in pounds volatile solids added per day per 1,000 cu ft if a digester 40 ft in diameter with a liquid level of 22 ft receives 28,500 gpd of sludge with 5.5% solids and 72% volatile solids? Assume the sludge weighs 8.34 lb/gal.

13.440 A digester 50 ft in diameter, with a liquid level of 20 ft, receives 36,220 gpd of sludge with 5.6% solids and 68% volatile solids. What is the digester loading in pounds volatile solids added per day per 1,000 cu ft? Assume the sludge weighs 8.34 lb/gal.

13.441 A digester 50 ft in diameter, with a liquid level of 18 ft, receives 16,200 gpd sludge with 5.1% solids and 72% volatile solids. What is the digester loading in the volatile solids added per day per 1,000 cu ft?

13.442 A total of 2,600 gpd sludge is pumped to a digester. If the sludge has a total solids content of 5.7% and a volatile solids concentration of 66%%, how many pounds of digested sludge should be in the digester for this load? Assume the sludge weighs 8.34 lb/gal. Use a ratio of 1 lb volatile solids added per day per 10 lb of digested sludge.

13.443 A total of 6,300 gpd of sludge is pumped to a digester. The sludge has a solids concentration of 5% and a volatile solids content of 70%. How many pound of digested sludge should be in the digester for this load? Assume the sludge weighs 8.34 lb/gal. Use a ratio of 1 lb volatile solids added per day per 10 lb of digested sludge.

13.444 The sludge pumped to a digester has a solids concentration of 6.5% and a volatile solids content of 67%. If a total of 5,200 gpd of sludge is pumped to the digester, how many pounds of digested sludge should be in the digester for this load? Assume the sludge weighs 8.34 lb/gal. Use a ratio of 1 lb volatile solids added per day per 10 lb of digested sludge.

13.445 A digester receives a flow of 3,800 gal of sludge during a 24 hr period. If the sludge has a solids content of 6% and a volatile solids concentration of 72%, how many pounds of digested sludge should be in the digester for this load? Assume the sludge weighs 8.34 lb/gal. Use a ratio of 1 lb volatile solids added per day per 10 lb of digested sludge.

13.446 The volatile acid concentration of the sludge in the anaerobic digester is 174 mg/L. If the measured alkalinity is 2,220 mg/L, what is the volatile acids/alkalinity ratio?

13.447 The volatile acid concentration of the sludge in the anaerobic digester is 160 mg/L. If the measured alkalinity is 2,510 mg/L, what is the volatility acids/alkalinity ratio?

13.448 The measured alkalinity is 2,410 mg/L. If the volatile acid concentration of the sludge in the anaerobic digester is 144 mg/L, what is the volatile acids/alkalinity ratio?

13.449 The measured alkalinity is 2,620 mg/L. If the volatile acid concentration of the sludge in the anaerobic digester is 178 mg/L, what is the volatile acids/alkalinity ratio?

13.450 To neutralize a sour digester, 1 mg/L of lime is to be added for every milligram per liter of volatile acids in the digester sludge. If the digester contains 244,000 gal of sludge with a volatile acid level of 2,280 mg/L, how many pounds of lime should be added?

13.451 To neutralize a sour digester, 1 mg/L of lime is to be added for every milligrams per liter of volatile acids in the digester sludge. If the digester contains 200,000 gal of sludge with a volatile acid level of 2,010 mg/L, how many pounds of lime should be added?

13.452 The digester contains 234,000 gal of sludge with a volatile acid level of 2,540 mg/L. To neutralize a sour digester, 1 mg/L of lime is to be added for every milligram per liter of volatile acids in the digester sludge. How many pounds of lime should be added?

13.453 The digester sludge is found to have a volatile acids content of 2,410 mg/L. If the digester volume is 182,000 gal, how many pounds of lime will be required for neutralization?

13.454 The sludge entering a digester has a volatile solids content of 68%. The sludge leaving the digester has a volatile solids content of 52%. What is the percent volatile solids reduction?

13.455 The sludge leaving the digester has a volatile solids content of 54%. The sludge entering a digester has a volatile solids content of 70%. What is the percent volatile solids reduction?

13.456 The raw sludge to a digester has a volatile solids content of 70%. The digested sludge volatile solids content is 53%. What is the percent volatile solids reduction?

13.457 The digested sludge volatile solids content is 54%. The raw sludge to a digester has a volatile solids content of 69%. What is the percent volatile solids reduction?

13.458 A flow of 3,800 gpd sludge is pumped to a 36,500 cu ft digester. The solids concentration of the sludge is 6.3%, with a volatile solids content of 73%. If the volatile solids reduction during digestion is 57%, how many pounds per day volatile solids are destroyed per cubic foot of digester capacity? Assume the sludge weighs 8.34 lb/gal.

13.459 A flow of 4,520 gpd sludge is pumped to a 33,000 cu ft digester. The solids concentration of the sludge is 7%, with a volatile solids content of 69%. If the volatile solids reduction during digestion is 54%, how many pounds per day volatile solids are destroyed per cubic foot of digester capacity? Assume the sludge weighs 8.34 lb/gal.

13.460 A 50 ft diameter digester receives a sludge flow of 2,600 gpd with a solids content of 5.6% and a volatile solids concentration of 72%. The volatile solids reduction during digestion is 52%. The digester operates at a level of 18 ft. What is the pounds per day volatile solids reduction per cubic foot of digester capacity? Assume the sludge weighs 8.34 lb/gal.

13.461 The sludge flow to a 40 ft diameter digester is 2,800 gpd, with a solids concentration of 6.1% and a volatile solids concentration of 65%. The digester is operated at a depth of 17 ft. If the volatile solids reduction during digestion is 56%, what is the pounds per day volatile solids reduction per 1,000 cu ft of digester capacity? Assume the sludge weighs 8.34 lb/gal.

13.462 A digester gas meter reading indicates an average of 6,600 cu ft of gas is produced per day. If a total of 500 lb/day volatile solids is destroyed, what is the digester gas production cubic foot gas per pound volatile solids destroyed?

13.463 A total of 2,110 lb of volatile solids is pumped to the digester daily. If the percent reduction of volatile solids due to digestion is 59% and the average gas production of the day is 19,330 cu ft, what is the daily gas production in cubic foot per pound volatile solids destroyed?

13.464 The total of 582 lb/day volatile solids is destroyed. If a digester gas meter reading indicates an average of 8,710 cu ft of gas is produced per day, what is the digester gas production in cubic foot per pound volatile solids destroyed?

13.465 The percent reduction of volatile solids due to digestion is 54%, and the average gas production for the day is 26,100 cu ft. If a total of 3,320 lb of volatile solids is pumped to the digester daily, what is the daily gas production in cubic foot per pound volatile solids destroyed?

13.466 A 40 ft diameter aerobic digester has a side water depth of 12 ft. The sludge flow to the digester is 9,100 gpd. Calculate the hydraulic digestion time in days.

13.467 An aerobic digester 40 ft in diameter has a side water depth of 10 ft. The sludge flow to the digester is 8,250 gpd. Calculate the hydraulic digestion time in days.

13.468 An aerobic digester is 80 ft long, 25 ft wide and has a side water depth of 12 ft. If the sludge flow to the digester is 7,800 gpd, what is the hydraulic digestion time in days?

13.469 A sludge flow of 11,000 gpd has a solids content of 3.4%. As a result of thickening, the sludge flow is reduced to 5,400 gpd, with a 5% solids content. Compare the digestion times for the two different sludge flows to a digester 30 ft in diameter with a side water depth of 12 ft.

13.470 An aerobic digester is 70 ft in diameter, with a side water depth of 10 ft. If the desired air supply for this digester was determined to be 40 cfm/1,000 cu ft digester capacity, what is the total cubic feet per minute air required for this digester?

13.471 The dissolved air concentrations recorded during a 5 min test of an air-saturated sample of aerobic digester sludge are given in the following table. Calculate the oxygen uptake in milligrams per liter per hour.

Elapsed Time, Min	DO Mg/L	Elapsed Time, Min	DO, Mg/L
At start	6.5	3 min	4.5
1 min	5.9	4 min	3.9
2 min	5.4	5 min	3.4

13.472 The dissolved air concentrations recorded during a 5 min test of an air-saturated sample of aerobic digester sludge are given in the following. Calculate the oxygen uptake in milligrams per liter per hour.

Elapsed Time, Min	DO, Mg/L	Elapsed Time, Min	DO, Mg/L
At start	7.3	3 min	4.3
1 min	6.8	4 min	4.2
2 min	5.7	5 min	3.6

13.473 Jar tests indicate that 22 mg of caustic are required to raise the pH of the 1 L sludge sample to 6.8. If the digester volume is 106,000 gal, how many pounds of caustic will be required for pH adjustment?

13.474 Jar tests indicate that 16 mg of caustic are required to raise the pH of the 1 L sludge sample to 6.8. If the digester volume is 148,000 gal, how many pounds of caustic will be required for pH adjustment?

13.475 A 2 L sample of digester sludge is used to determine the required caustic dosage for pH adjustment. If 64 mg of caustic are required for pH adjustment in the jar test and the digester volume is 54,000 gal, how many pounds of caustic will be required for pH adjustment?

13.476 A 2 L sample of digested sludge is used to determine the required dosage for pH adjustment. A total of 90 mg caustic is used in the jar test. The aerobic digester is 60 ft in diameter, with a side water depth of 14 ft. How many pounds of caustic are required for pH adjustment of the digester?

13.477 Sludge is being pumped to the digester at a rate of 3.6 gpm. How many pounds of volatile solids are being pumped to the digester daily if the sludge has 5.1% total solids content, with 71% volatile solids?

13.478 A 55 ft diameter anaerobic digester has a liquid depth of 22 ft. The unit receives 47,200 gal of sludge daily, with a solids content of 5.3%, of which 71% is volatile. What is the organic loading rate in the digester in pounds volatile solids added per cubic foot per day?

13.479 The concentration of volatile acids in the anaerobic digester is 181 mg/L. If the concentration of alkalinity is measured to be 2,120 mg/L, what is the volatile acids/alkalinity ratio?

13.480 If the anaerobic digester becomes sour, it must be neutralized. Adding lime to the unit can do this. The amount of lime to add is determined by the ratio of 1 mg/L of lime for every milligram per liter of volatile acids in the digester. If the volume of sludge in the digester is 756,000 L and the volatile acids concentration is 1,820 mg/L, how many kilograms of lime will be required to neutralize the digester?

13.481 The anaerobic digester has a raw sludge volatile solids content of 67%. The digested sludge has a volatile solids content of 55%. What is the percent reduction in the volatile solids content through the anaerobic digester?

13.482 Calculations indicate that 2,600 kg of volatile solids will be required in the seed sludge. How many liters of seed sludge will be required if the sludge has a 9.5% solids content with 66% volatile solids and weighs 1.14 kg/L?

13.483 A total of 8,200 gpd sludge is pumped to a digester. If the sludge has a solids content of 5.7% and a volatile solids concentration of 65%, how many pounds of digested sludge should be in the digester for this load? Use a ratio of 1 lb volatile solids per day per 10 lb of digested sludge.

13.484 If 4,400 lb/day solids with a volatile solids content of 67% are sent to the digester, how many pounds of volatile solids are sent to the digester daily?

13.485 What is the digester loading in pound volatile solids added per day per 1,000 cu ft if a digester 60 ft in diameter, with a liquid level of 20 ft, receives 12,900 gpd of sludge with 5.4% solids and 65% volatile solids?

13.486 A primary sludge flow of 4,040 gpd with a solids content of 5.4% is mixed with a thickened secondary sludge flow of 5,820 gpd with a solids content of 3.3%. What is the percent solids content of the mixed sludge flow? Assume both sludges weigh 8.34 lb/gal?

13.487 A sludge pump has a bore of 8 in and a stroke length of 6 in. The counter indicates a total of 3,500 revolutions during a 24 hr period. What is the pumping rate in gallons per day? Assume 100% efficiency.

13.488 A 60 ft diameter digester has a typical side water depth of 24 ft. If 88,200 gal seed sludge are to be used in starting up the digester, what percent of the digester volume will be seed sludge?

13.489 A flow of 3,800 gpd sludge is pumped to a 36,000 cu ft digester. The solids content of the sludge is 4.1%, with a volatile solids content of 70%. If the volatile solids reduction during digestion is 54%, how many pounds per day volatile solids are destroyed per cubic foot of digester capacity? Assume the sludge weighs 8.34 lb/gal.

13.490 The volatile acid concentration sludge in the anaerobic digester is 156 mg/L. If the measured alkalinity is 2,310 mg/L, what is the volatile acids/alkalinity ratio?

13.491 To neutralize a sour digester, 1 mg/L of lime is to be added for every milligram per liter of volatile acid in the digester sludge. If the digester contains 240,000 gal of sludge with a volatile acid level of 2,240 mg/L, how many pounds of lime should be added?

13.492 A 50 ft diameter digester has a typical water depth of 22 ft. If the seed sludge to be used is 24% of the tank capacity, how may gallons of seed sludge will be required?

13.493 A total of 4,310 gpd of sludge is to be pumped to the digester. If the sludge has 5.3% solids content with 72% volatile solids, how many pounds per day volatile solids are pumped to the digester? Assume the sludge weighs 8.34 lb/gallon.

13.494 Primary and thickened secondary sludges are to be mixed and sent to the digester. The 2,940 gpd primary sludge has a solids content of 5.9%, and the 4,720 gpd thickened secondary sludge has a solids content of 3.8%. What would be the percent solids content of the mixed sludge? Assume both sludges weigh 8.34 lb/gal.

13.495 The measured alkalinity is 2,470 mg/L. If the volatile acid concentration of the sludge in the anaerobic digester is 150 mg/L, what is the volatile acids/alkalinity ratio?

13.496 A total of 42,250 lb/day sludge is pumped to a 94,000 gal digester. The sludge being pumped to the digester has total solids content of 4% and volatile solids content of 60%. The sludge in the digester has a solids content of 6.0%, with a 55% volatile solids content. What is the volatile solids loading on the digester in pounds volatile solids added per day per pound volatile solids in digester?

13.497 A sludge pump has a bore of 9 in and a stroke length of 5 in. If the pump operates at 30 strokes per minute, how many gallons per minute are pumped? Assume 100% efficiency.

13.498 What is the digester loading in pound volatile solids added per day per 1,000 cu ft if a digester 40 ft in diameter, with a liquid level of 21 ft, receives 19,200 gpd of sludge with 5% solids and 66% volatile solids?

13.499 The digester sludge is found to have a volatile acids content of 2,200 mg/L. If the digester volume is 0.3 MG, how many pounds of lime will be required for neutralization?

13.500 A digester gas meter reading indicates an average of 6,760 cu ft of gas is produced per day. If a total of 580 lb/day volatile solids is destroyed, what is the digester gas production in cubic foot gas per pound volatile solids destroyed?

13.501 The sludge entering a digester has a volatile solids content of 67%. The sludge leaving the digester has a volatile solids content of 52%. What is the percent volatile solids reduction?

13.502 The raw sludge flow to a new digester is expected to be 1,230 gpd. The raw sludge contains 4.1% solids and 66% volatile solids. The desired volatile solids loading ratio is 0.09 lb volatile solids added per pound volatile solids in the digester. How many gallons of seed sludge will be required if the seed sludge contains 7.5% solids with a 55% volatile solids content? Assume the raw sludge weighs 8.34 lb/gal and the seed sludge weighs 8.5 lb/gallon.

13.503 The sludge leaving the digester has a volatile solids content of 56%. The sludge entering a digester has a volatile solids content of 70%. What is the percent volatile solids retention?

13.504 A 60 ft diameter aerobic digester has a side water depth of 12 ft. The sludge flow to the digester is 9,350 gpd. Calculate the digestion time in days.

13.505 A total of 2,610 lb of volatile solids is pumped to the digester daily. If the percent reduction of volatile solids due to digestion is 56% and the average gas production for the day is 22,400 cu ft, what is the daily gas production in cubic foot per pound volatile solids destroyed?

13.506 The sludge flow to a 50 ft diameter digester is 3,200 gpd, with a solids content of 6.4% and a volatile solids concentration of 68%. The digester is operated at a depth of 22 ft. If the volatile solids reduction during digestion is 55%, what is the pounds per day volatile solids reduction per 1,000 cu ft of digester capacity?

13.507 The desired air supply rate for an aerobic digester is determined to be 0.05 cfm/cu ft digester capacity. What is the total cubic feet per minute air required if the digester is 80 ft long, 20 ft wide and has a side water depth of 12 ft?

13.508 Jar testing indicates that 22 mg of caustic are required to raise the pH of the 1 L sample to 7. If the digester volume is 120,000 gal, how many pounds of caustic will be required for pH adjustment?

13.509 Dissolved air concentration is taken on an air-saturated sample of digested sludge at 1 min intervals. Given the results that follow, calculate the oxygen uptake in milligrams per liter per hour.

Elapsed Time, Min	DO Mg/L	Elapsed Time, Min	DO, Mg/L
At start	7.7	3 min	5.2
1 min	6.9	4 min	4.5
2 min	6.0	5 min	3.8

13.510 The flow to a primary clarifier is 2.2 MGD. The influent suspended solids concentration is 220 mg/L, and the effluent suspended solids concentration is 101 mg/L. If the sludge to be removed from the clarifier has solids content of 3.0% and the sludge pumping rate is 25 gpm, how many minutes per hour should the pump operate?

13.511 A sludge flow of 12,000 gpd has a solids concentration of 2.6%. The solids concentration is increased to 4.6% as a result of thickening, and the reduced flow rate is 5,400 gpd. Compare the digestion time for these two different sludge flows. The digester is 32 ft in diameter, with a 24 ft operating depth.

13.512 A filter press used to dewater digested primary sludge receives a flow of 1,100 gal during a 3 hr period. The sludge has a solids content of 3.8%. If the plate surface area is 140 sq ft, what is the solids loading rate in pounds per hour per square foot? Assume the sludge weighs 8.34 lb/gal.

13.513 A filter press used to dewater digested primary sludge receives a flow of 820 gal during a 2 hr period. The solids content of the sludge is 5%. If the plate surface area is 160 sq ft, what is the solids loading rate in pounds per hour per square foot? Assume the sludge weighs 8.34 lb/gal.

13.514 A plate and frame filter press receives solids loading of 0.80 lb/hr/sq ft. If the filtration time is 2 hr and the time required to remove the sludge cake and being sludge fed to the press is 20 min, what is the net filter yield in pounds per hour per square foot?

13.515 A plate and frame filer press receives a flow of 680 gal of sludge during a 2 hr period. The solids concentration of the sludge is 3.9%. The surface area of the plate is 130 sq ft. If the downtime for sludge cake discharge is 20 min, what is the net filter yield in pounds per hour per square foot? Assume the sludge weighs 8.34 lb/gal.

13.516 A 6 ft wide belt press receives a flow of 140 gpm of primary sludge. What is the hydraulic loading rate in gallons per minute per foot?

13.517 The amount of sludge to be dewatered by the belt filter press is 21,300 lb/day. If the belt filter press is to be operated 12 hr each day, what should the pounds per hour sludge feed rate be to the press?

13.518 The amount of sludge to be dewatered by a belt filter press is 23,100 lb/day. If the maximum feed rate that still provides an acceptable cake is 1,800 lb/hr, how many hours per day should the belt remain in operation?

13.519 The sludge fed to a belt filter press is 160 gpm. If the total suspended solids concentration of the feed is 4.4%, what is the solids loading rate in pounds per hour? Assume the sludge weighs 8.34 lb/gal.

13.520 The flocculant concentration for a belt filter press is 0.7%. If the flocculant feed rate is 4 gpm, what is the flocculation feed rate in pounds per hour? Assume the flow is steady and continuous, and assume the flocculant weighs 8.34 lb/gallon.

13.521 Digested sludge is applied to a vacuum filter at a rate of 80 gpm, with a solids concentration of 5.1%. If the vacuum filter has a surface area of 320 sq ft, what is the filter loading in pounds per hour per square foot? Assume the sludge weighs 8.34 lb/gal.

13.522 The wet cake flow from a vacuum filter is 6,810 lb/hr. If the filter area is 320 sq ft and the percent solids in the cake is 31%, what is the filter yield in pounds per hour per square foot?

13.523 A total of 5,400 lb/day primary sludge solids is to be processed by a vacuum filter. The vacuum filter yield is 3.3 lb/hr/sq ft. The solids recovery is 90%. If the area of the filter is 230 sq ft, how many hours per day must the vacuum filter remain in operation to process this many solids?

13.524 The total pounds of dry solids pumped to a vacuum filter during a 24 hr period is 18,310 lb/day. The vacuum filter is operated 10 hr/day. If the percent solids recovery is 91% and the filter area is 265 sq ft, what is the filter yield in pounds per hour per square foot?

13.525 The sludge fed to a vacuum filter is 85,230 lb/hr, with a solids content of 5.9%. If the wet cake flow is 18,400 lb/hr, with 20% solids content, what is the percent solids content? What is the percent solids recovery?

13.526 A drying bed is 210 ft long and 22 ft wide. If sludge is applied to a depth of 8 in, how many gallons of sludge are applied to the drying bed?

13.527 A drying bed is 240 ft long and 26 ft wide. If sludge is applied to a depth of 8 in, how many gallons of sludge are applied to the drying beds?

13.528 A sludge bed is 190 ft long and 20 ft wide. A total of 168,000 lb of sludge is applied each application of the sand drying bed. The sludge has a solids content of 4.6%. If the drying-and-removal cycle requires 21 days, what is the solids loading rate in pounds per year per square foot?

13.529 A sludge drying bed is 220 ft long and 30 ft wide. The sludge is applied to a depth of 9 in. The solids concentration of the sludge is 3.9%. If the drying-and-removal cycle requires 25 days, what is the solids loading rate to the beds in pounds per year per square foot? Assume the sludge weighs 8.34 lb/gallon.

13.530 Sludge is withdrawn from a digester which has a diameter of 50 ft. If the sludge is drawn down 2.4 ft, how many cubic foot will be sent to the drying beds?

13.531 A 50 ft diameter digester has a drawdown of 14 in. If the drying bed is 70 ft long and 40 ft wide, how many feet deep will the drying be as a result of the drawdown?

13.532 If 4,700 lb/day dewatered sludge, with a solids content of 21%, are mixed with 3,800 lb/day compost, with a 26% moisture content, what is the percent moisture of the blend?

13.533 The total dewatered digested primary sludge produced at a plant is 4,800 lb/day, with a solids content of 17%. The final compost to be used in blending has moisture content of 27%. How much compost (pounds per day) must be blended with the dewatered sludge to produce a mixture with moisture content of 42%?

13.534 Compost is blended from bulking material and dewatered sludge. The bulking material is to be mixed with 7.4 cu yds of dewatered sludge at a ratio (by volume) of 3:1. The solids content of the sludge is 19%, and the solids content of the bulking material is 54%. If the bulk density of the sludge is 1,710 lb/cu yd and the bulk density of the bulking material is 760 lb/cu yd, what is the percent solids of the compost blend?

13.535 A composting facility has an available capacity of 8,200 cu yds. If the composting cycle is 21 days, how many pounds per day wet compost can be processed by this facility? Assume a compost bulk density of 1,100 lb/cu yd.

13.536 Compost is to be blended from wood chips and dewatered sludge. The wood chips are to be mixed with 12 cu yds of dewatered sludge at a ratio (by volume) of 3:1. The solids content of the sludge is 16%, and the solids content of the wood chips is 55%. If the bulk density of the sludge is 1,720 lb/cu yd and the bulk density of the wood chips is 820 lb/cu yd, what is the percent solids of the compost blend?

13.537 Given the data listed in the following, calculate the solids processing capability, in pounds per day, of the compost operation.

Cycle time: 21 days
Total available capacity: 7,810 cu yds
Percent solids of wet sludge: 19%
Mix ratio (by volume) of wood chips to sludge: 3
Wet compost bulk density: 1,100 lb/cu yd
Wet sludge bulk density: 1,720 lb/cu yd
Wet wood chips bulk density: 780 lb/cu yd

13.538 The sludge fed to a belt filter press is 150 gpm. If the total suspended solids concentration of the feed is 4.8%, what is the solids loading rate in pounds per hour? Assume the sludge weighs 8.34 lb/gallon.

13.539 Sludge is applied to a drying bed 220 ft long and 24 ft wide. The sludge has a total solids concentration of 3.3% and fills the bed to a depth of 10 in. If it takes an average of 22 days for the sludge to dry and 1 day to remove the dried solids, how many pounds of solids can be dried for every square foot of drying bed area each year?

13.540 A belt filter press receives a daily sludge flow of 0.20 million gallons. If the belt is 70 in wide, what is the hydraulic loading rate on the unit in gallons per minute for each foot of belt width (gpm/ft)?

13.541 A plate and frame filter press can process 960 gal of sludge during its 140 min operating cycle. If the sludge concentration is 4.2%, and if the plate surface area is 150 sq ft, how many pounds of solids are pressed per hour for each square foot of plate surface area?

13.542 Thickened thermally conditioned sludge is pumped to a vacuum filter at a rate of 36 gpm. The vacuum area of the filter is 10 ft wide, with a drum diameter of 9.6 ft. If the sludge concentration is 12%, what is the filter yield in pounds per hour per square foot? Assume the sludge weight 8.34 lb/gallon.

13.543 The vacuum filter produces an average of 3,020 pounds of sludge cake each hour. The total solids content of the cake produced is 40%. The sludge is being pumped to the filter at a rate of 24 gpm and at a concentration of 11%. If the sludge density is 8.50 lb/gallon, what is the percent recovery of the filter?

13.544 The amount of sludge to be dewatered by the belt press is 25,200 lb/day. If the belt filter press is to be operated 12 hr each day, what should be the sludge feed rate in pounds per hour to the press?

13.545 A filter press used to dewater digested primary sludge receives a flow of 800 gal of sludge during a 2 hr period. The sludge has solids content of 4.1%. If the plate surface area is 141 sq ft, what is the solids loading rate in pounds per hour per square foot? Assume the sludge weighs 8.34 lb/gallon.

13.546 The sludge feed rate to belt filter press is 170 gpm. The total suspended solids concentration of the feed is 5%. The flocculation used for sludge conditioning is a 0.9% concentration, with a feed rate of 2.8 gpm. What is the flocculant dose expected as pound flocculant per ton of solids treated?

13.547 A plate and frame filter press receives solids loading of 0.8 lb/hr/sq ft. If the filtration time is 2 hr and the time required to remove the sludge cake and begin sludge feed to the press is 20 min, what is the net filter yield in pounds per hour per square foot? Assume the sludge weighs 8.34 lb/gallon.

13.548 Laboratory tests indicate that the total residue portion of a feed sludge sample is 24,300 mg/L. The total filterable residue is 740 mg/L. On this basis, what is the estimated total suspended solids concentration of the sludge sample?

13.549 Digested sludge is applied to a vacuum filter at a rate of 80 gpm, with a solids concentration of 5.5%. If the vacuum filter has a surface area of 320 sq ft, what is the filter loading in pounds per hour per square foot? Assume the sludge weighs 8.34 lb/gallon.

13.550 The wet cake flow from a vacuum filter is 7,500 lb/hr. If the filter area is 320 sq ft and the percent solids in the cake is 26%, what is the filter yield in pounds per hour per square foot?

13.551 The amount of sludge to be dewatered by a belt filter is 28,300 lb/day. If the maximum feed rate which still provides an acceptable cake is 1,800 lb/hr, how many hours per day should the belt remain in operation?

13.552 A total of 5,700 lb/day primary sludge solids is to be processed by a vacuum filter. The vacuum filter yield is 3.1 lb/hr/sq ft. The solids recovery is 92%. If the area of the filter is 280 sq ft, how many hours per day must the vacuum filter remain in operation to process this many solids?

13.553 A drying bed is 220 ft long and 30 ft wide. If sludge is applied to a depth of 9 in, how many gallons of sludge are applied to the drying bed?

13.554 The sludge fed to a vacuum filter is 91,000 lb/day, with a solids content of 5.3%. If the wet cake flow is 14,300 lb/day, with a 28% solids content, what is the percent solids recovery?

13.555 A sludge drying bed is 200 ft long and 25 ft wide. The sludge is applied to a depth of 8 in. The solids concentration of the sludge is 5.1%. If the drying-and-removal cycle requires 20 days, what is the solids loading rate to the beds in pound per year per square foot? Assume the sludge weighs 8.34 lb/gallon.

13.556 A drying bed is 190 ft long and 30 ft wide. If a 40 ft diameter digester has a drawdown of 1 ft, how many feet deep will the drying bed be as a result of the drawdown?

13.557 A treatment plant produces a total of 6,800 lb/day of dewatered digested primary sludge. The dewatered sludge has a solids concentration 25%. Final compost to be used in blending has moisture content of 36%. How much compost (in pounds per day) must be blended with the dewatered sludge to produce a mixture with a moisture content of 55%?

13.558 Compost is to be blended from wood chips and dewatered sludge. The wood chips are to be mixed with 7.0 cu yds of dewatered sludge at a ratio of 3:1. The solids content of the sludge is 16%, and the solids content of the wood chips is 51%. If the bulk density of the sludge is 1,710 lb/cu yd and the bulk density of the wood chips is 780 lb/cu yd, what is the percent solids of the compost blend?

13.559 A composting facility has an available capacity of 6,350 cu yds. If the composting cycle is 26 days, how many pounds per day wet compost can be processed by this facility? How many tons per day is this? Assume a compost bulk density of 980 lb/cu yd.

13.560 Given the data listed in the following, calculate the dry sludge processing capability, in pounds per day, of the compost operation.

Cycle time: 24 days
Total available capacity: 9,000 cu yds
Percent solids of wet sludge: 18%
Mix ratio (by volume) of wood chips to sludge: 3.1
Wet compost bulk density: 1,100 lb/cu yd
Wet sludge bulk density: 1,710 lb/cu yd
Wet wood chips bulk density: 800 lb/cu yd

13.1 ADDITIONAL WATER/WASTEWATER PRACTICE PROBLEMS

Tank Volume Calculations

1. The diameter of a tank is 70 ft. If the water depth is 25 ft, what is the volume of water in the tank in gallons?

2. A tank is 60 ft in length, 20 ft wide, and 10 ft deep. Calculate the cubic feet volume of the tank.

3. A tank 20 ft wide and 60 ft long is filled with water to a depth of 12 ft. What is the volume of the water in the tank (in gallons)?

4. What is the volume of water in a tank, in gallons, if the tank is 20 ft wide, 40 ft long and contains water to a depth of 12 ft?

5. **A tank has a diameter of 60 ft and a depth of 12 ft. Calculate the volume of water in the tank in gallons.**

6. What is the volume of water in a tank, in gallons, if the tank is 20 ft wide, 50 ft long and contains water to a depth of 16 ft?

Channel and Pipeline Capacity Calculations

7. A rectangular channel is 340 ft in length, 4 ft in depth, and 6 ft wide. What is the volume of water in cubic feet?

8. A replacement section of 10 in pipe is to be sandblasted before it is put into service. If the length of the pipeline is 1,600 ft, how many gallons of water will be needed to fill the pipeline?

9. A trapezoidal channel is 800 ft in length, 10 ft wide at the top, 5 ft wide at the bottom, with a distance of 4 ft from top edge to bottom along the sides. Calculate the gallon volume.

10. A section of 8 in diameter pipeline is to be filled with treated water for distribution. If the pipeline is 2,250 ft in length, how many gallons of water will be distributed?

11. A channel is 1,200 ft in length, carries water 4 ft in depth, and is 5 ft wide. What is the volume of water in gallons?

Miscellaneous Volume Calculations

12. A pipe trench is to be excavated that is 4 ft wide, 4 ft deep, and 1,200 ft long. What is the volume of the trench in cubic yards?

13. A trench is to be excavated that is 3 ft wide, 4 ft deep, and 500 yd long. What is the cubic yard volume of the trench?

14. A trench is 300 yd long, 3 ft wide, and 3 ft deep. What is the cubic feet volume of the trench?

15. A rectangular trench is 700 ft long, 6.5 ft wide, and 3.5 ft deep. What is the cubic feet volume of the trench?

16. The diameter of a tank is 90 ft. If the water depth in the tank is 25 ft, what is the volume of water in the tank in gallons?

17. A tank is 80 ft long, 20 ft wide, and 16 ft deep. What is the cubic feet volume of the tank?

18. How many gallons of water will it take to fill an 8 in diameter pipe that is 4,000 ft in length?

19. A trench is 400 yd long, 3 ft wide, and 3 ft deep. What is the cubic feet volume of the trench?

20. A trench is to be excavated. If the trench is 3 ft wide, 4 ft deep, and 1,200 ft long, what is the cubic yard volume of the trench?

21. A tank is 30 ft wide and 80 ft long. If the tank contains water to a depth of 12 ft, how many gallons of water are in the tank?

22. What is the volume of water (in gallons) contained in a 3,000 ft section of channel if the channel is 8 ft wide and the water depth is 3.5 ft?

23. A tank has a diameter of 70 ft and a depth of 19 ft. What is the volume of water in the tank in gallons?

24. If a tank is 25 ft in diameter and 30 ft deep, how many gallons of water will it hold?

Flow, Velocity, and Conversion Calculations

25. A channel 44 in wide has water flowing to a depth of 2.4 ft. If the velocity of the water is 2.5 fps, what is the cubic feet per minute flow in the channel?

26. A tank is 20 ft long and 12 ft wide. With the discharge valve closed, the influent to the tank causes the water level to rise 0.8 ft in 1 min. What is the gallons per minute flow to the tank?

27. A trapezoidal channel is 4 ft wide at the bottom and 6 ft wide at the water surface. The water depth is 40 in. If the flow velocity through the channel is 130 ft/min, what is the cubic feet per minute flow rate through the channel?

28. An 8 in diameter pipeline has water flowing at a velocity of 2.4 fps. What is the gallons per minute flow rate through the pipeline? Assume the pipe is flowing full.

29. A pump discharges into a 3 ft diameter container. If the water level in the container rises 28 in in 30 sec, what is the gallons per minute flow into the container?

30. A 10 in diameter pipeline has water flowing at a velocity of 3.1 fps. What is the gallons per minute flow rate through the pipeline if the water is flowing at a depth of 5 in?

31. A channel has a rectangular cross section. The channel is 6 ft wide, with water flowing to a depth of 2.6 ft. If the flow rate through the channel is 14,200 gpm, what is the velocity of the water in the channel (in feet per second)?

32. An 8 in diameter pipe flowing full delivers 584 gpm. What is the velocity of flow in the pipeline (in feet per second)?

33. A special dye is used to estimate the velocity of flow in an interceptor line. The dye is injected into the water at one pumping station, and the travel time to the first manhole 550 ft away is noted. The dye first appears at the downstream manhole in 195 sec. The dye continues to be visible until the total elapsed time is 221 sec. What is the feet per second velocity of flow through the pipeline?

34. The velocity in a 10 in diameter pipeline is 2.4 ft/sec. If the 10 in pipeline flows into an 18 in diameter pipeline, what is the velocity in the 8 in pipeline in feet per second?

35. A float travels 500 ft in a channel in 1 min, 32 sec. What is the estimated velocity in the channel (in feet per second)?

36. The velocity in an 8 in diameter pipe is 3.2 ft/sec. If the flow then travels through a 10 in diameter section of pipeline, what is the feet per second velocity in the 10 in pipeline?

Average Flow Rates

37. The following flows were recorded for the week:

 Monday: 4.8 MGD
 Tuesday: 5.1 MGD
 Wednesday: 5.2 MGD
 Thursday: 5.4 MGD

Friday: 4.8 MGD
Saturday: 5.2 MGD
Sunday: 4.8 MGD
What was the average daily flow rate for the week?

38. The totalizer reading the month of September was 121.4 MG. What was the average daily flow (ADF) for the month of September?

Flow Conversions

39. Convert 0.165 MGD to gallons per minute.

40. The total flow for 1 day at a plant was 3,335,000 gal. What was the average gallons per minute flow for that day?

41. Express a flow of 8 cfs in terms of gallons per minute.

42. What is 35 gps expressed as gallons per day?

43. Convert a flow of 4,570,000 gpd to cubic feet per minute.

44. What is 6.6 MGD expressed as cubic feet per second?

45. Express 445,875 cfd as gallons per minute.

46. Convert 2,450 gpm to gallons per day.

General Flow and Velocity Calculations

47. A channel has a rectangular cross section. The channel is 6 ft wide, with water flowing to a depth of 2.5 ft. If the flow rate through the channel is 14,800 gpm, what is the velocity of the water in the channel (in feet per second)?

48. A channel 55 in wide has water flowing to a depth of 3.4 ft. If the velocity of the water is 3.6 fps, what is the cubic feet per minute flow in the channel?

49. The following flows were recorded for the months of June, July, and August: June, 102.4 MG; July, 126.8 MG; August, 144.4 MG. What was the average daily flow for this three-month period?

50. A tank is 12 ft by 12 ft. With the discharge valve closed, the influent to the tank causes the water level to rise 8 in in 1 min. What is the gallons per minute flow to the tank?

51. An 8 in diameter pipe flowing full delivers 510 gpm. What is the feet per second velocity of flow in the pipeline?

52. Express a flow of 10 cfs in terms of gallons per minute.

53. The totalizer reading for the month of December was 134.6 MG. What was the average daily flow (ADF) for the month of September?

54. What is 5.2 MGD expressed as cubic feet per second?

55. A pump discharges into a 3 ft diameter container. If the water level in the container rises 20 in in 30 sec, what is the gallons per minute flow into the container?

56. Convert a flow of 1,825,000 gpd to cubic feet per minute.

57. A 6 in diameter pipeline has water flowing at a velocity of 2.9 fps. What is the gallons per minute flow rate through the pipeline?

58. The velocity in a 10 in pipeline is 2.6 ft/sec. If the 10 in pipeline flows into an 8 in diameter pipeline, what is the foot per second velocity in the 8 in pipeline?

59. Convert 2,225 gpm to gallons per day.

60. The total flow for 1 day at a plant was 5,350,000 gal. What was the average gallons per minute flow for that day?

Chemical Dosage Calculations

61. Determine the chlorinator setting (in pounds per day) needed to treat a flow of 5.5 MGD with a chlorine dose of 2.5 mg/L.

62. To dechlorinate a wastewater, sulfur dioxide is to be applied at a level 4 mg/L more than the chlorine residual. What should the sulfonator feed rate be (in pounds per day) for a flow of 4.2 MGD with a chlorine residual of 3.1 mg/L?

63. What should the chlorinator setting be (in pounds per day) to treat a flow of 4.8 MGD if the chlorine demand is 8.8 mg/L and a chlorine residual of 3 mg/L is desired?

64. A total chlorine dosage of 10 mg/L is required to treat the water in a unit process. If the flow is 1.8 MGD and the hypochlorite has 65% available chlorine, how many pounds per day of hypochlorite will be required?

65. The chlorine dosage at a plant is 5.2 mg/L. If the flow rate is 6,250,000 gpd, what is the chlorine feed rate in pounds per day?

66. A storage tank is to be disinfected with 60 mg/L of chlorine. If the tank holds 86,000 gal, how many pounds of chlorine (gas) will be needed?

67. To neutralize a sour digester, 1 lb of lime is to be added for every pound of volatile acids in the digester liquor. If the digester contains 225,000 gal of sludge with a volatile acid (VA) level of 2,220 mg/L, how many pounds of lime should be added?

68. A flow of 0.83 MGD requires a chlorine dosage of 8 mg/L. If the hypochlorite has 65% available chlorine, how many pounds per day of hypochlorite will be required?

13.1.1 Advanced Practice Problems

1. The diameter of a tank is 60 ft. If the water depth is 20 ft, what is the volume of water in the tank in gallons?
2. A tank is 50 ft in length, 15 ft wide, and 10 ft deep. Calculate the cubic feet volume of the tank.
3. A tank 10 ft wide and 50 ft long is filled with water to a depth of 10 ft. What is the volume of the water in the tank (in gallons)?
4. What is the volume of water in a tank, in gallons, if the tank is 10 ft wide, 40 ft long and contains water to a depth of 10 ft?

5. A tank has a diameter of 50 ft and a depth of 12 ft. Calculate the volume of water in the tank in gallons.
6. What is the volume of water in a tank, in gallons, if the tank is 20 ft wide, 50 ft long and contains water to a depth of 16 ft?
7. A rectangular channel is 300 ft in length, 4 ft in depth, and 6 ft wide. What is the volume of water in cubic feet?
8. A replacement section of 10 in pipe is to be sandblasted before it is put into service. If the length of pipeline is 1,500 ft, how many gallons of water will be needed to fill the pipeline?
9. A trapezoidal channel is 700 ft in length, 10 ft wide at the top, 5 ft wide at the bottom, with a distance of 4 ft from top edge to bottom along the sides. Calculate the gallon volume.
10. A section of 8 in diameter pipeline is to be filled with treated water for distribution. If the pipeline is 2,100 ft in length, how many gallons of water will be distributed?
11. A channel is 1,100 ft in length, carries water 4 ft in depth, and is 5 ft wide. What is the volume of water in gallons?
12. A pipe trench is to be excavated that is 4 ft wide, 4 ft deep, and 1,100 ft long. What is the volume of the trench in cubic yards?
13. A trench is to be excavated that is 3 ft wide, 4 ft deep, and 400 yd long. What is the cubic yard volume of the trench?
14. A trench is 270 yd long, 3 ft wide, and 3 ft deep. What is the cubic feet volume of the trench?
15. A rectangular trench is 600 ft long, 6.5 ft wide, and 3.5 ft deep. What is the cubic feet volume of the trench?
16. The diameter of a tank is 90 ft. If the water depth in the tank is 20 ft, what is the volume of water in the tank in gallons?
17. A tank is 80 ft long, 20 ft wide, and 12 ft deep. What is the cubic feet volume of the tank?
18. How many gallons of water will it take to fill an 8 in diameter pipe that is 3,000 ft in length?
19. A trench is 500 yd long, 3 ft wide, and 3 ft deep. What is the cubic feet volume of the trench?
20. A trench is to be excavated. If the trench is 3 ft wide, 4 ft deep, and 1,100 ft long, what is the cubic yard volume of the trench?
21. A tank is 30 ft wide and 60 ft long. If the tank contains water to a depth of 12 ft, how many gallons of water are in the tank?
22. What is the volume of water (in gallons) contained in a 2,000 ft section of channel if the channel is 8 ft wide and the water depth is 3.5 ft?
23. A tank has a diameter of 60 ft and a depth of 19 ft. What is the volume of water in the tank in gallons?
24. If a tank is 20 ft in diameter and 30 ft deep, how many gallons of water will it hold?
25. A channel 44 in wide has water flowing to a depth of 2.4 ft. If the velocity of the water is 2.0 fps, what is the cubic feet per minute flow in the channel?

26. A tank is 20 ft long and 12 ft wide. With the discharge valve closed, the influent to the tank causes the water level to rise 0.7 ft in 1 min. What is the gallons per minute flow to the tank?
27. A trapezoidal channel is 4 ft wide at the bottom and 6 ft wide at the water surface. The water depth is 40 in. If the flow velocity through the channel is 120 ft/min, what is the cubic feet per minute flow rate through the channel?
28. An 8 in diameter pipeline has water flowing at a velocity of 2.2 fps. What is the gallons per minute flow rate through the pipeline? Assume the pipe is flowing full.
29. A pump discharges into a 2 ft diameter container. If the water level in the container rises 28 in in 30 sec, what is the gallons per minute flow into the container?
30. A 10 in diameter pipeline has water flowing at a velocity of 3.0 fps. What is the gallons per minute flow rate through the pipeline if the water is flowing at a depth of 5 in?
31. A channel has a rectangular cross section. The channel is 6 ft wide, with water flowing to a depth of 2.5 ft. If the flow rate through the channel is 14,200 gpm, what is the velocity of the water in the channel (in feet per second)?
32. An 8 in diameter pipe flowing full delivers 590 gpm. What is the velocity of flow in the pipeline (foot per second)?
33. The velocity in a 10 in diameter pipeline is 2.4 ft/sec. If the 10 in pipeline flows into an 18 in diameter pipeline, what is the velocity in the 8 in pipeline in feet per second?
34. A float travels 400 ft in a channel in 1 min, 32 sec. What is the estimated velocity in the channel (in feet per second)?
35. The velocity in an 8 in diameter pipe is 3.2 ft/sec. If the flow then travels through a 10 in diameter section of pipeline, what is the feet per second velocity in the 10 in pipeline?
35. The following flows were recorded for the week:

    Monday: 4.8 MGD
    Tuesday: 5.0 MGD
    Wednesday: 5.2 MGD
    Thursday: 5.4 MGD
    Friday: 4.8 MGD
    Saturday: 5.2 MGD
    Sunday: 4.8 MGD
    What was the average daily flow rate for the week?

37. The totalizer reading the month of September was 124.4 MG. What was the average daily flow (ADF) for the month of September?
38. Convert 0.175 MGD to gallons per minute.
39. The total flow for 1 day at a plant was 3,330,000 gal. What was the average gallons per minute flow for that day?

40. Express a flow of 7 cfs in terms of gallons per minute.
41. What is 30 gps expressed as gallons per day?
42. Convert a flow of 4,500,000 gpd to cubic feet per minute.
43. What is 6.5 MGD expressed as cubic feet per second?
44. Express 445,875 cfd as gallons per minute.
45. Convert 2,450 gpm to gallons per day.
46. A channel has a rectangular cross section. The channel is 6 ft wide, with water flowing to a depth of 2 ft. If the flow rate through the channel is 14,800 gpm, what is the velocity of the water in the channel (in feet per second)?
47. A channel 55 in wide has water flowing to a depth of 3.4 ft. If the velocity of the water is 3.5 fps, what is the cubic feet per minute flow in the channel?
48. The following flows were recorded for the months of June, July, and August: June, 107.4 MG; July, 126.8 MG; August, 144.4 MG. What was the average daily flow for this three-month period?
49. A tank is 10 ft by 10 ft. With the discharge valve closed, the influent to the tank causes the water level to rise 8 in in 1 min. What is the gallons per minute flow to the tank?
50. An 8 in diameter pipe flowing full delivers 510 gpm. What is the foot per second velocity of flow in the pipeline?
51. Express a flow of 11 cfs in terms of gallons per minute.
52. The totalizer reading for the month of December was 134.6 MG. What was the average daily flow (ADF) for the month of September?
53. What is 5.0 MGD expressed as cubic feet per second?
54. A pump discharges into a 3 ft diameter container. If the water level in the container rises 20 in in 30 sec, what is the gallons per minute flow into the container?
55. Convert a flow of 1,820,000 gpd to cubic feet per minute.
56. A 6 inch diameter pipeline has water flowing at a velocity of 2.9 fps. What is the gallons per minute flow rate through the pipeline?
57. The velocity in a 10 in pipeline is 2.6 ft/sec. If the 10 in pipeline flows into an 8 in diameter pipeline, what is the feet per second velocity in the 8 in pipeline?
58. Convert 2,220 gpm to gallons per day.
59. The total flow for 1 day at a plant was 5,300,000 gal. What was the average gallons per minute flow for that day?

Appendix A: Answers Chapter 13, Wastewater Treatment Practice Problems

13.1 (4.9 MGD) (1 day/24 hrs) (96 hrs) = 19.6 MG

$$\frac{(4.3\text{ ft})\ (5.8\text{ ft})\ (28\text{ in.})}{19.6\text{ MG}} \times \frac{1\text{ ft}}{12\text{ in.}} = 3.0\text{ cu ft/MG}$$

13.2
$$\frac{(700\text{ gpm})}{(7.48\text{ gal/cu ft}} = 93.6\text{ cu ft/min}$$

$$\frac{93.6\text{ cu ft}}{1\text{ minute}} \times \frac{1\text{ minute}}{60\text{ seconds}} \times \frac{1}{(1.4\text{ ft} \times 1.7\text{ ft})} = 0.7\text{ ft/second}$$

13.3
$$\frac{1{,}200{,}000\text{-gal}}{1\text{ day}} \times \frac{1\text{ day}}{24\text{ hrs}} \times \frac{1\text{ hr}}{60\text{ min}}$$

$$= \frac{1{,}200{,}000\text{ gal}}{1440\text{ min/day}} = 833\text{ gpm}$$

$$= \frac{16\text{ in.}}{12\text{ in.}} = 1.4\text{ ft} \frac{18\text{ in.}}{12\text{ in}} = 1.6\text{ ft}$$

$$= \frac{883\text{-gal}}{1\text{ minute}} \times \frac{1\text{ cu ft}}{7.48\text{ gal}} \times \frac{1\text{ minute}}{60\text{ seconds}} = 1.97\text{ cu ft/seconds}$$

$$= \frac{1.97\text{ cu ft}}{1\text{ seconds}} \times \frac{1}{1.33\text{ ft} \times 1.5\text{ ft}} = 0.99\text{ ft/second}$$

13.4 (12 ft) (10 ft) (6 ft) (7.48 gal/cu ft) = 5,386 gal

13.5 (14 ft) (12 ft) (6 ft) (7.48 gal/cu ft) = 7,540 gal

13.6 (9 ft) (9 ft) (6 ft) = 486 cu ft

13.7 (12 ft) (8 ft) (x ft) (7.48 gal/cu ft) = 4,850 gal

$x = 6.8$ ft

13.8 (10 ft) (8 ft) (3.1 ft) (7.48 gal/cu ft) = 1,855 gal

13.9
$$\frac{(10\text{ ft})\ (10\text{ ft})\ (1.8\text{ ft})\ (7.48\text{ gal/cu ft})}{5\text{ minutes}} = 269\text{ gpm}$$

13.10
$$\frac{(12\text{ ft})\ (12\text{ ft})\ (1.3\text{ ft})\ (7.48\text{ gal/cu ft})}{3\text{ minutes}} = 467\text{ gpm}$$

13.11 $$\frac{(9\text{ ft})\ (7\text{ ft})\ (1.4\text{ ft})\ (7.48\text{ gal/cu ft})}{3\text{ min}} = 220\text{ gpm}$$

13.12 $$\frac{(9\text{ ft})\ (8\text{ ft})\ (0.9\text{ ft})\ (7.48\text{ gal/cu ft})}{1\text{ min}} = 485\text{ gpm}$$

13.13 (12 ft) (14 ft) (–2.2 ft) (7.48 gal/cu ft) = –2,765 gal

$$\frac{-2765\text{ gal}}{5\text{ min}} = -553\text{ gpm}$$

0 gpm = discharge (+) (–553 gpm)
533 gpm = discharge

13.14 $$\text{Accumulation} = \frac{(12\text{ ft})\ (14\text{ ft})\ (-1.9\text{ ft})\ (7.48\text{ gal/cu ft})}{4\text{ min}}$$

= –597 gpm
0 gpm = discharge + (–597 gpm)
597 gpm – discharge

13.15 10 ft, 9 in. = 10.75 ft
12 ft – 2 in = 12.17 ft

$$\text{Accumulation} = \frac{(10.75\text{ ft})\ (12.12\text{ ft})\ (-2\text{ ft})\ (7.48\text{ gal/cu ft})}{5.5\text{ min}} = -354\text{ gpm}$$

0 gpm = discharge + (–354 gpm)
discharge = 354 gpm

13.16 410 gpm = discharge + accumulation

$$\frac{(11.8\text{ ft})\ (14\text{ ft})\ (0.083\text{ ft})\ (7.48\text{ gal/cu ft})}{7} = 14.5\text{ gpm}$$

410 gpm = discharge + 145 gpm
discharge = 395.5 gpm

13.17 140 in = 11.7 ft
148 in = 12.3 ft

$$\text{Accumulation} = \frac{(11.7\text{ ft})\ (12.3\text{ ft})\ (-0.125\text{ ft})\ (7.48\text{ gal/cu ft})}{6\text{ min}}$$

= –22.4 gpm
430 gpm = discharge + (–22.4 gpm)
discharge = 452.4 gpm

13.18 $$\text{Total Accumulation} = \frac{(9.8\text{ ft})\ (14\text{ ft})\ (-0.7\text{ ft})\ (7.48\text{ gal/cu ft})}{15\text{ min}} = -48\text{ gpm}$$

800 gpm = discharge + (–48 gpm)
848 gpm = discharge
848 gpm = 500 gpm + second pump
348 gpm = second pump discharge

13.19 $$\frac{60\text{ gpd}}{7.48\text{ gal/cu ft}} = 8\text{ cu ft/day}$$

13.20 $$\frac{282\text{ gal/wk}}{(7.48\text{ gal/cu ft})\ (7\text{ days/wk})} = 5.4\text{ cu ft/day}$$

13.21 $\frac{4.9 \text{ cu ft}}{3.33 \text{ MGD}} = 1.5 \text{ cu ft/MG}$

13.22 $\frac{\frac{81 \text{ gal/day}}{7.48 \text{ gal/cu ft}}}{4.9 \text{ MGD}} = 2.2 \text{ cu ft/MG}$

13.23 $\frac{\frac{48 \text{ gal/day}}{7.48 \text{ gal/cu ft}}}{2.28 \text{ MGD}} = 2.8 \text{ cu ft/MG}$

13.24 $\frac{600 \text{ cu ft}}{2.9 \text{ cu ft/day}} = 207 \text{ days}$

13.25 $\frac{(9 \text{ cu yds})(27 \text{ cu ft/cu yd})}{1.6 \text{ cu ft/day}} = 152 \text{ days}$

13.26 (2.6 cu ft/MG) (2.9 MGD) = 7.5 cu ft/day

$\frac{292 \text{ cu ft}}{7.5 \text{ cu ft/day}} = 39 \text{ days}$

13.27 $\frac{x \text{ cu ft}}{3.5 \text{ cu ft/day}} = 120 \text{ days}$

$x = 420$ cu ft

13.28 $(4 \text{ ft})(1.6 \text{ ft})(\text{fps}) = \frac{1820 \text{ gpm}}{(7.48 \text{ gal/cu ft})(60 \text{ sec/min})}$

$x = 0.6$ fps

13.29 26 ft/32 sec = 0.8 ft/sec

13.30 (3 ft) (1.2 ft) (ft/sec) = 4.3 cfs

$x = 1.2$ fps

13.31 36 ft/32 sec = 1.2 ft/sec

13.32 $(2.8 \text{ ft})(1.3 \text{ ft})(x \text{ fps}) = \frac{1140 \text{ gpm}}{(7.48 \text{ gal/cu ft})(60 \text{ sec/min})}$

$x = 0.7$ fps

13.33 $\frac{12 \text{ cu ft/day}}{8 \text{ MGD}} = 1.5 \text{ cu ft/MG}$

13.34 $\frac{260 \text{ gal/day}}{(7.48 \text{ gal/cu ft})(11.4 \text{ MGD})} = 3 \text{ cu ft/MG}$

13.35 (3.1 cu ft/MG) (3.8 MGD) = 11.8 cu ft/day

(11.8 cu ft/day) (30 days/month) = 354 cu ft/month

$\frac{354 \text{ cu ft/month}}{27 \text{ cu ft/cu yd}} 13.1 \text{ cu yds/month}$

13.36 2.2 cu ft/MG (4.23 MGD) (90 days) = 838 cu ft required

$\frac{838 \text{ cu ft}}{27 \text{ cu ft/cu yd}} = 13 \text{ cu yds}$

13.37 (2.6 ft) (1.3 ft) (1.1 fps) = 3.7 cfs

13.38 (3 ft) (1.2 ft) (1.4 fps) (7.48 gal/cu ft) (60 sec/min) (1,440 min/day) = 3,257,211 gpd

13.39 (2.7 ft) (0.83 ft) (0.90 fps) = 2 cfs

13.40

$$\begin{array}{r} 25.6782\text{ g} \\ -25.6662\text{ g} \\ \hline .0120\text{ g} \end{array}$$

0.0120 g × 1,000 mg/1 = 12 mg/0.05L = 240 mg/L

13.41

$$\begin{array}{r} 26.2410\text{ g} \\ -26.2345\text{ g} \\ \hline 0.0065\text{ g} \end{array}$$

(0.0065 g) (1,000 mg) = 6.5 mg

$$\frac{6.5\text{ mg}}{0.026\text{ L}} = 250\text{ mg/L}$$

13.42

$$\begin{array}{r} 8.42\text{ mg/L} \\ -4.28\text{ mg/L} \\ \hline 4.14\text{ mg/L} \end{array}$$

$$\frac{6\text{ mL}}{300\text{ mL}} = 0.02$$

$$\frac{4.14\text{ mg/L}}{0.02} = 207\text{ mg/L}$$

13.43 BOD = (7.96 mg/L – 4.26 mg/L) × 300 mL/5 mL = 222 mg/L

13.44 (14.6 mg/L) (3.13 MGD) (8.34 lb/gal) = 381 lb/day

13.45 (310 mg/L) (6.15 MGD) (8.34 lb/gal) = 15,900 lb/day

13.46 (188 mg/L) (4.85 MGD) (8.34 lb/gal) = 7,604 lb/day

13.47 $$\frac{270\text{ cu ft}}{(2.6\text{ cu ft/MG})\ (2.95\text{ MGD})} = 35\text{ days}$$

13.48 $$\frac{210\text{ gal}}{(7.48\text{ gal/cu ft})\ (7\text{ days})} = 4\text{ cu ft/day}$$

13.49 $$\frac{5.4\text{ cu ft}}{2.91\text{ MGD}} = 1.85\text{ cu ft/MG}$$

13.50 $$\frac{(12\text{ cu yds})\ (27\text{ cu ft/cu yd})}{2.4\text{ cu ft/day}} = 135\text{ days}$$

13.51 $$\frac{36\text{ ft}}{31\text{ seconds}} = 1.2\text{ ft/sec}$$

13.52 (26 ft) (1.3 ft) (0.8 fps) = 2.7 cfs

13.53 $$\frac{210 \text{ gallons}}{(7.48 \text{ gal/cu ft})\ (8.8 \text{ MGD})} = 3.2 \text{ cu ft/MG}$$

13.54 (2.6 ft) (1.3 ft) (1.8 ft/sec) (7.48 gal/cu ft) (60 sec/min) = 2,730 gal

13.55 2.3 cu ft/MG (3.61 MGD) (30 days) = 249 cu ft

$$\frac{249 \text{ cu ft}}{27 \text{ cu ft/cu yd}} = 9.2 \text{ cu yds}$$

13.56 (3 ft) (0.83 ft) (1 fps) = 2.5 cfs

13.57 $$\frac{160{,}000 \text{ gal}}{75{,}417 \text{ gph}} = 2.1 \text{ hrs}$$

13.58 $$\frac{(90 \text{ ft})\ (25 \text{ ft})\ (10 \text{ ft})\ (7.48 \text{ gal/cu ft})}{135{,}416 \text{ gph}} = 1.2 \text{ hrs}$$

13.59 $$\frac{(0.785)\ (90 \text{ ft})\ (90 \text{ ft})\ (12 \text{ ft})\ (7.48 \text{ gal/cu ft})}{181{,}250 \text{ gph}}$$

$$\frac{570{,}739}{181{,}250 \text{ gph}} = 3.1 \text{ hrs}$$

13.60 (84 ft) (20 ft) (13.1 ft) (7.48 gal/cu ft) = 164,619 gal

$$\text{Detention Time} = (164{,}619 \text{ gal})\left(\frac{1 \text{ day}}{2{,}010{,}000 \text{ gal}}\right)(24 \text{ hr/day}) = 1.97 \text{ hr}$$

13.61 (90 ft) (20 ft) = 1,800 sq ft

$$= \frac{1{,}450{,}000 \text{ gpd}}{1800 \text{ sq ft}}$$

$$= 806 \text{ gpd/sq ft}$$

13.62

73.86 g	23.10 g
−22.20 g	−22.20 g
----------	------------
51.66 g (Sample weight)	0.90 g (Dry solids weight)

$$\frac{0.90 \text{ g}}{51.66 \text{ g}} \times 100\% = 1.7\%$$

13.63 (390 gal/min) (60 min/hr) (24 hrs/day) (8.34 lb/gal) (0.8%) = 37,470 lb/day

13.64 140 mg/L − 50 mg/L = 90 mg/L

$$\% \text{ Removal} = \frac{90 \text{ mg/L}}{140 \text{ mg/L}} \times 100 = 64.2\%$$

13.65 $$\frac{1{,}420{,}000 \text{ gpd}}{80 \text{ ft}} = 17{,}750 \text{ gpd/ft}$$

13.66 (80 ft) (20 ft) (12 ft) (7.48 gal/cu ft) = 143,616 gal

$$\text{Detention Time} = (3\text{ tanks})\ (143{,}616\text{ gal})\frac{1}{5{,}000{,}000\text{ gal}}(24\text{ hr/day})$$
$$(60\text{ min/hr}) = 124\text{ min}$$

$$\text{Surface Overflow Rate} = \frac{5{,}000{,}000\text{ gpd}}{(3)\ (80\text{ft})\ (20\text{ ft})} = 1{,}042\text{ gpd/sq ft}$$

$$\text{Weir Overflow Rate} = \frac{5{,}000{,}000\text{ gpd}}{(3)\ (86\text{ ft})} = 19{,}380\text{ gpd/ft}$$

13.67 $$\frac{(80\text{ ft})\ (35\text{ ft})\ (12\text{ ft})\ (7.48\text{ gal/cu ft})}{135{,}000\text{ gph}} = 1.9\text{ hrs}$$

13.68 $$\frac{1{,}520{,}000\text{ gpd}}{112\text{ ft}} = 13{,}571\text{ gpd/ft}$$

13.69 $$\frac{2{,}980{,}000\text{ gpd}}{(3.14)\ (70\text{ ft})} = 13{,}558\text{ gpd/ft}$$

13.70 $$\frac{(2520\text{ gpm})\ (1440\text{ min/day})}{(3.14)\ (90\text{ ft})} = 12{,}841\text{ gpd/ft}$$

13.71 $$\frac{1{,}880{,}000\text{ gpd}}{192\text{ ft}} = 9792\text{ gpd/ft}$$

13.72 $$\frac{2{,}910{,}000\text{ gpd}}{(0.785)\ (70\text{ ft})\ (70\text{ ft})} = 756\text{ gpd/sq ft}$$

13.73 $$\frac{2{,}350{,}000\text{ gpd}}{(80\text{ ft})\ (30\text{ ft})} = 979\text{ gpd/sq ft}$$

13.74 $$\frac{2{,}620{,}000\text{ gpd}}{(80\text{ ft})\ (30\text{ ft})} = 1092\text{ gpd/sq ft}$$

13.75 $$\frac{(2610\text{ gpm})\ (1440\text{ min/day})}{(0.785)\ (60\text{ ft})\ (60\text{ ft})} = 1330\text{ gpd/sq ft}$$

13.76 $$\frac{(3110\text{ mg/L})\ (4.1\text{ MGD})\ (8.34\text{ lb/gal})}{(0.785)\ (70\text{ ft})\ (70\text{ ft})} = 27.6\text{ lb/day/sq ft}$$

13.77 $$\frac{(3220\text{ mg/L})\ (4.4\text{ MGD})\ (8.34\text{ lb/gal})}{(0.785)\ (80\text{ ft})\ (80\text{ ft})} = 23.5\text{ lb/day/sq ft}$$

13.78 $$\frac{(2710\text{ mg/L})\ (3.22\text{ MGD})\ (8.34\text{ lb/gal})}{(0.785)\ (70\text{ ft})\ (70\text{ ft})} = 18.9\text{ lb/day/sq ft}$$

13.79 $$\frac{(3310\text{ mg/L})\ (2.98\text{ MGD})\ (8.34\text{ lb/gal})}{(0.785)\ (80\text{ ft})\ (80\text{ ft})} = 16.4\text{ lb/day/sq ft}$$

13.80 (125 mg/L) (5.55 MGD) (8.34 lb/gal) = 5,786 lb/day SS

13.81 (40 mg/L rem.) (2.92 MGD) (8.34 lb/gal) = 974 lb/day SS

13.82 (90 mg/L rem.) (4.44 MGD) (8.34 lb/gal) = 3,333 lb/day BOD

13.83 (200 mg/L rem.) (0.98 MGD) (8.34 lb/gal) = 1,635 lb/day SS

13.84 $$\frac{135 \text{ mg/L Rem.}}{220 \text{ mg/L}} \times 100 = 61\% \text{ Removal Efficiency}$$

13.85 $$\frac{111 \text{ mg/L}}{188 \text{ mg/L}} \times 100 = 59\% \text{ Removal Efficiency}$$

13.86 $$\frac{220 \text{ mg/L Rem.}}{280 \text{ mg/L}} \times 100 = 79\% \text{ Removal}$$

13.87 $$\frac{111 \text{ mg/L}}{300 \text{ mg/l}} \times 100 = 37\% \text{ Removal Efficiency}$$

13.88 $$\frac{(0.785)\,(80 \text{ ft})\,(80 \text{ ft})\,(10 \text{ ft})\,(7.48 \text{ gal/cu ft})}{171{,}667 \text{ gph}} = 2.2 \text{ hr}$$

13.89 $$\frac{2{,}320{,}000 \text{ gpd}}{(0.785)\,(60 \text{ ft})\,(60 \text{ ft})} = 821 \text{ gpd/sq ft}$$

13.90 $$\frac{3{,}728{,}000 \text{ gpd}}{215 \text{ ft}} = 17{,}340 \text{ gpd/sq ft}$$

13.91 $$\frac{(2710 \text{ mg/L})\,(2.46 \text{ MGD})\,(8.34 \text{ lb/gal})}{(0.785)\,(60 \text{ ft})\,(60 \text{ ft})} = 19.7 \text{ lb/day/sq ft}$$

13.92 $$\frac{3{,}100{,}000 \text{ gpd}}{(0.785)\,(70 \text{ ft})\,(70 \text{ ft})} = 806 \text{ gpd/sq ft}$$

13.93 $$\frac{(2910 \text{ mg/L})\,(3.96 \text{ MGD})\,(8.34 \text{ lb/gal})}{(0.785)\,(80 \text{ ft})\,(80 \text{ ft})} = 19.1 \text{ lb/day/sq ft}$$

13.94 (118 mg/L rem.) (5.3 MGD) (8.34 lb/gal) = 5,216 lb/day rem.

13.95 $$\frac{(90 \text{ ft})\,(40 \text{ ft})\,(14 \text{ ft})\,(7.48 \text{ gal/cu ft})}{212{,}500 \text{ gph}} = 1.8 \text{ hr}$$

13.96 $$\frac{(1940 \text{ gpm})\,(1440 \text{ min/day})}{(3.14)\,(70 \text{ ft})} = 12{,}710 \text{ gpd/ft}$$

13.97 (84 mg/L) (4.44 MGD) (8.34 lb/gal) = 3,110 lb/day BOD removed

13.98 (210 mg/L rem.) (3.88 MGD) (8.34 lb/gal) = 6,795 lb/day removed

13.99 $$\frac{191 \text{ mg/L Removed}}{260 \text{ mg/L}} \times 100 = 73\%$$

13.100 $$\frac{2{,}220{,}000 \text{ gpd}}{(90 \text{ ft})\,(40 \text{ ft})} = 617 \text{ gpd/sq ft}$$

13.101 $$\frac{780{,}000 \text{ gpd}}{(0.785)\,(80 \text{ ft})\,(80 \text{ ft})} = 155 \text{ gpd/sq ft}$$

13.102 $$\frac{(2360\text{ gpm})\ (1440\text{ min/day})}{(0.985)\ (90\text{ ft})\ (90\text{ ft})} = 534\text{ gpd/sq ft}$$

13.103 $$\frac{1{,}500{,}000\text{ gpd}}{(0.785)\ (90\text{ ft})\ (90\text{ ft})} = 236\text{ gpd/sq ft}$$

13.104 $$\frac{(0.785)\ (96\text{ ft})\ (96\text{ ft})}{43{,}560\text{ sq ft/ac}} = 0.17\text{ ac}$$

$$\frac{2.1\text{ MGD}}{0.17\text{ ac}} = 12.4\text{ MGD/ac}$$

13.105 $$\frac{(210\text{ mg/L})\ (1.4\text{ MGD})\ (8.34\text{ lb/gal})}{47.1\times 1000\text{-cu ft}} = 52\text{ lb BOD/1000 cu ft}$$

13.106 $$\frac{(111\text{ mg/L})\ (3.40\text{ MGD})\ (8.34\text{ lb/gal})}{44.5\times 1000\text{-cu ft}} = 70.7\text{ lb BOD/day/1000 cu ft}$$

13.107 $$\frac{(201\text{ mg/L})\ (0.9\text{ MGD})\ (8.34\text{ lb/gal})}{35.1\times 1000\text{-cu ft}} = 43\text{ lb BOD/day/1000 cu ft}$$

13.108 $$\frac{(0.785)\ (90\text{ ft})\ (90\text{ ft})\ (5\text{ ft})}{43{,}560\text{ cu ft/ac-ft}} = 0.73\text{ ac-ft}$$

$$\frac{(120\text{ mg/L})\ (1.4\text{ MGD})\ (8.34\text{ lb/gal})}{0.73\text{ ac-ft}} = 1919\text{ lb BOD/day/ac-ft}$$

13.109 (122 mg/L rem.) (3.24 MGD) (8.34 lb/gal) = 3,297 lb/day SS rem.

13.110 (176 mg/L rem.) (1.82 MGD) (8.34 lb/gal) = 2,671 lb/day BOD rem.

13.111 (182 mg/L BOD rem.) (2.92 MGD) (8.34 lb/gal) = 4,432 lb/day BOD rem.

13.112 (194 mg/L BOD rem.) (5.4 MGD) (8.34 lb/gal) = 8,737 lb/day BOD rem.

13.113 $$\frac{101\text{ mg/L Removed}}{149\text{ mg/L}} \times 100 = 68\%$$

13.114 $$\frac{239\text{ mg/L Removed}}{261\text{ mg/L}} \times 100 = 92\%\text{ Removal Efficiency}$$

13.115 $$\frac{179\text{ mg/L Removed}}{201\text{ mg/L}} \times 100 = 89\%\text{ Removal Efficiency}$$

13.116 $$\frac{88\text{ mg/L Removed}}{111\text{ mg/L}} \times 100 = 79\%$$

13.117 $$\frac{3.5\text{ MGD}}{3.4\text{ MGD}} = 1$$

13.118 $$\frac{2.32\text{ MGD}}{1.64\text{ MGD}} = 1.4$$

13.119 $\frac{3.86 \text{ MGD}}{2.71 \text{ MGD}} = 1.4$

13.120 $1.6 = \frac{\text{x MGD}}{4.6 \text{ MGD}}$

$x = 7.4$ MGD

13.121 0.310 MGD + 0.355 MGD = 0.655 MGD
surface area = (0.785) (90 ft) (90 ft) = 6,359 sq ft

$\frac{655{,}000 \text{ gpd}}{6359 \text{ sq ft}} = 103 \text{ gpd/sq ft}$

13.122 $\frac{4.55 \text{ MGD}}{2.8 \text{ MGD}} = 1.6$

13.123 (75 mg/L) (1.35 MGD) (8.34 lb/gal) = 844.4 lb BOD/day
surface area = (0.785) (80 sq ft) (80 sq ft) = 5,024 sq ft
(5,024 sq ft) (6 ft) = 30,144 cu ft

$\frac{(844.4 \text{ lb day})\ (1000 \text{ lb BOD})}{30{,}144} = 28 \text{ lb BOD/d/1000 cu ft}$

13.124 81 mg/L − 13 mg/L = 68 mg/L removed
(68 mg/L) (4.1 MGD) (8.34 lb/gal) = 2,325 BOD/day

13.125 $\frac{630{,}000 \text{ gpd}}{(0.785)\ (80 \text{ ft})\ (80 \text{ ft})} = 125 \text{ gpd/sq ft}$

13.126 (114 mg/L rem.) (2.84 MGD) (8.34 lb/day) = 2,700 lb/day SS rem.

13.127 $\frac{131 \text{ mg/L Rem.}}{200 \text{ mg/L}} \times 100 = 65.5\%$

13.128 (156 mg/L rem.) (1.44 MGD) (8.34 lb/gal) = 1,873 lb/day rem.

13.129 $\frac{2{,}880{,}000 \text{ gpd}}{(0.785)\ (90 \text{ ft})\ (90 \text{ ft})} = 453 \text{ gpd/sq ft}$

13.130 $\frac{188 \text{ mg/L Rem.}}{210 \text{ mg/L}} \times 100 = 90\%$ Removal Efficiency

13.131 $\frac{(0.785)\ (90 \text{ ft})\ (90 \text{ ft})\ (6 \text{ ft})}{43{,}560 \text{ cu ft/ac-ft}} = 0.9 \text{ ac-ft}$

$\frac{(144 \text{ mg/L})\ (1.26 \text{ MGD})\ (8.34 \text{ lb/gal})}{0.9 \text{ ac-ft}} = 1681 \text{ BOD/day/ac-ft}$

13.132 (178 mg/L rem.) (4.22 MGD) (8.34 lb/gal) = 6,265 lb/day rem.

13.133 $\frac{3.8 \text{ MGD}}{3.6 \text{ MGD}} = 1.1$

13.134 $\frac{(0.785)\ (80 \text{ ft})\ (80 \text{ ft})}{43{,}560 \text{ sq ft/ac}} = 0.12 \text{ ac}$

$$\frac{1.9 \text{ MGD}}{0.12 \text{ ac}} = 15.8 \text{ MGD/ac}$$

13.135 $$\frac{1{,}930{,}00 \text{ gpd}}{(0.785)\ (90 \text{ ft})\ (90 \text{ ft})} = 304 \text{ gpd/sq ft}$$

13.136 $$\frac{(166 \text{ mg/L})\ (0.81 \text{ MGD})\ (8.34 \text{ lb/gal})}{23.1 \times 1000\text{-cu ft}} = 48 \text{ lb BOD/day/1000 cu ft}$$

13.137 $$\frac{2.35 \text{ MGD}}{1.67 \text{ MGD}} = 1.4$$

13.138 $$\frac{208 \text{ mg/L Rem.}}{243 \text{ mg/L}} \times 100 = 86\% \text{ Removal Efficiency}$$

13.139 $$\frac{2{,}980{,}000 \text{ gpd}}{720{,}000 \text{ sq ft}} = 4.1 \text{ gpd/sq ft}$$

13.140 $$\frac{4{,}725{,}000 \text{ gpd}}{880{,}000 \text{ sq ft}} = 5.4 \text{ gpd/sq ft}$$

13.141 $$\frac{1{,}550{,}000 \text{ gpd}}{440{,}000 \text{ sq ft}} = 3.5 \text{ gpd/sq ft}$$

13.142 $$\frac{x \text{ gpd}}{800{,}000 \text{ sq ft}} = 7 \text{ gpd/sq ft}$$

x = 5,600,000 gpd

13.143 (241 mg/L SS) (0.55 K-value) = 133 mg/L particulate BOD

13.144 222 mg/L total BOD = (241 mg/L) (0.5 K-value) + x mg/L soluble BOD
x = 102 mg/L soluble BOD

13.145 240 mg/L total BOD = (150 mg/L) (0.5 K-value) + x mg/L soluble BOD
x = 165 mg/L soluble BOD

13.146 288 mg/L Total BOD = (268 mg/L) (0.6 K-value) + x mg/L soluble BOD
x = 127 mg/L soluble BOD
(127 mg/L) (1.9 MGD) (8.34 lb/gal) = 2,012 lb/day soluble BOD
14.147 187 mg/L total BOD = (144 mg/L) (0.52) + x mg/L soluble BOD
x = 112 mg/L Soluble BOD

$$\frac{(112 \text{ mg/L})\ (2.8 \text{ MGD})\ (8.34 \text{ lb/gal})}{765 \times 1000\text{-sq ft}} = 3.4 \text{ lb/day/1000 sq ft}$$

13.148 $$\frac{450{,}000 \text{ gpd}}{190{,}000 \text{ sq ft}} = 2.4 \text{ gpd/sq ft}$$

13.149 (190 mg/L – (0.6 × 210 mg/L) = 64 mg/L

13.150 $$\frac{(1.9 \text{ MGD})\ (128 \text{ mg/L})\ (8.34 \text{ lb/gal})}{410{,}000 \text{ sq ft}} \times 1000$$
= 4.9 lb/day/100 sq ft

13.151 210 mg/L – (0.65 × 240 mg/L) = 54 mg/L

$$\frac{(0.71\text{ MGD})\ (54\text{ mg/L})\ (8.34\text{ lb/gal})}{110{,}000\text{ sq ft}} \times 100$$

= 2.9 lb/day/1,000 sq ft

13.152 Hydraulic Loading

$$\frac{455{,}000\text{ gpd}}{206{,}000\text{ sq ft}} = 2.2\text{ gpd/sq ft}$$

Unit Organic Loading

(241 mg/L) – (0.5 × 149 mg/L) = 166.5 mg/L

(0.455 MGD) (166.5 mg/L) (8.34 lb/gal) = 632 lb/day

$$\frac{632\text{ lb/day}}{206{,}000\text{ sq ft}} 1000 = 3.1\text{ lb/day/1000 sq ft}$$

First Stage Organic Loading

$$\frac{632\text{ lb/day}}{(0.785)\ (103{,}000\text{ sq ft})} \times 1000 = 8.2\text{ lb/day/1000 sq ft}$$

$$\frac{632\text{ lb/day}}{103{,}000\text{ sq ft}} \times 1000 = 6.1\text{ lb/day/1000 sq ft}$$

13.153 $$\frac{2{,}960{,}000\text{ gpd}}{660{,}000\text{ sq ft}} = 4.5\text{ gpd/sq ft}$$

13.154 (222 mg/L) (0.5 K-value) = 111 mg/L

13.155 210 mg/L total BOD = (205 mg/L) (0.6) + x mg/L soluble BOD

x = 87 mg/L soluble BOD

(87 mg/L) (2.9 MGD) (8.34 lb/gal) = 2,104 lb/day

13.156 $$\frac{(121\text{ mg/L})\ (2.415\text{ MGD})\ (8.34\text{ lb/gal})}{760 \times 1000\text{ sq-ft}} = 3.2\text{ lb/day/1000 sq ft}$$

13.157 (80 ft) (30 ft) (14 ft) (7.48 gal/cu ft) = 251,328 gal

13.158 (80 ft) (30 ft) (12 ft) (7.48 gal/cu ft) = 215,424 gal

13.159 (0.785) (80 ft) (80 ft) (12 ft) (7.48 gal/cu ft) = 450,954 gal

13.160 (0.785) (70 ft) (70 ft) (10 ft) (7.48 gal/cu ft) = 287,718 gal

13.161 (240 mg/L) (0.88 MGD) (8.34 lb/gal) = 1,761 lb/day BOD

13.162 (160 mg/L) (4.29 MGD) (8.34 lb/gal) = 5,725 lb/day COD

13.163 (165 mg/L) (3.24 MGD) (8.34 lb/gal) = 4,459 lb/day BOD

13.164 (150 mg/L) (4.88 MGD) (8.34 lb/gal) = 6,105 lb/day BOD

13.165 (2,110 mg/L) (0.46 MG) (8.34 lb/gal) = 8,095 lb MLSS

13.166 (2,420 mg/L) (0.54 MG) (8.34 lb/gal) = 10,899 lb MLVSS

13.167 (2,410 mg/L) (0.38 MG) (8.34 lb/gal) = 7,638 lb MLVSS

13.168 (2,740 mg/L) (0.39 MG) (8.34 lb/gal) = 8,912 lb MLVSS

13.169 (2,470 mg/L) (0.66 MG) (8.34 lb/gal) (0.73) = 9,925 lb MLVSS

13.170 $$\frac{(198\text{ mg/L})\ (2.72\text{ MGD})\ (8.34\text{ lb/gal})}{(2610\text{ mg/L})\ (0.48\text{ MG})\ (8.34)} = 0.43$$

13.171 $$\frac{(148\text{ mg/L})\ (3.35\text{ MGD})\ (8.34\text{ lb/gal})}{(2510\text{ mg/L})\ (0.49\text{ MG})\ (8.34\text{ lb/gal})} = 0.40$$

13.172 $$\frac{(180\text{ mg/L})\ (0.32\text{ MGD})\ (8.34\text{ lb/gal})}{(2540\text{ mg/L})\ (0.195\text{ MG})\ (8.34\text{ lb/gal})} = 0.12$$

13.173 $$\frac{(181\text{ mg/L})\ (3.3\text{ MGD})\ (8.34\text{ lb/gal})}{\text{x lb MLVSS}} = 0.7$$

$x = 7{,}116\text{ mg/L MLVSS}$

13.174 $$\frac{(141\text{ mg/L})\ (2.51\text{ MGD})\ (8.34\text{ lb/gal})}{\text{x lb MLVSS}} = 0.4$$

$x = 7{,}379\text{ lb MLVSS}$

13.175 $$\frac{16{,}100\text{ lb MLVSS}}{2630\text{ lb/day}} = 6.1\text{ days}$$

13.176 $$\frac{(2720\text{ mg/L})\ (0.48\text{ MG})\ (8.34\text{ lb/gal})}{(110\text{ mg/L})\ (2.9\text{ MGD})\ (8.34\text{ lb/gal})} = 4.1\text{ days}$$

13.177 $$\frac{(2510\text{ mg/L})\ (0.57\text{ MGD})\ (8.34\text{ lb/gal})}{(111\text{ mg/L})\ (2.88\text{ MGD})\ (8.34\text{ lb/gal})} = 4.5\text{ days}$$

13.178 $$\frac{(2960\text{ mg/L})\ (0.58\text{ MGD})\ (8.34\text{ lb/gal})}{(110\text{ mg/L})\ (1.98\text{ MGD})\ (8.34\text{ lb/gal})} = 7.9\text{ days}$$

13.179 $$\frac{(3810\text{ mg/L})\ (0.211\text{ MG})\ (8.34\text{ lb/gal})}{(205\text{ mg/L})\ (0.27\text{ MGD})\ (8.34\text{ lb/gal})} = 14.6\text{ days}$$

13.180 $$\frac{29{,}100\text{ lb}}{2920\text{ lb/day} + 400\text{ lb/day}} = 8.8\text{ days}$$

13.181 $$\frac{(2710\text{ mg/L})\ (1.5\text{ MG})\ (8.34\text{ lb/gal}) + (1940\text{ mg/L})\ (0.106\text{ MG})\ (8.34\text{ lb/gal})}{(5870\text{ mg/L})\ (0.072\text{ MGD})\ (8.34\text{ lb/gal}) + (25\text{ mg/L})\ (3.3\text{ MGD})\ (8.34\text{ lb/gal})}$$

$$\frac{33{,}902\text{ lb} + 1715\text{ lb}}{3225\text{ lb/day} + 688\text{ lb/day}} = 9.1\text{ days}$$

13.182 $$\frac{(2222\text{ mg/L})\ (0.612\text{ MG})\ (8.34\text{ lb/gal})}{1610\text{ lb/day} + 240\text{ lb/day}} = 6.1\text{ days}$$

13.183 $$\frac{(2910\text{ mg/L})\ (0.475\text{ MG})\ (8.34\text{ lb/gal})}{(6210\text{ mg/L})\ (0.027\text{ MGD})\ (8.34\text{ lb/gal}) + (16\text{ mg/L})\ (1.4\text{ MGD})\ (8.34\text{ lb/gal})}$$

$$\frac{11{,}528}{1398\text{ lb/day} + 187\text{ lb/day}} = 7.3\text{ days}$$

13.184 $$\frac{220\text{ mL/L}}{1000\text{ mL/L} - 220\text{ mL/L}} = 0.28$$

13.185 $$\frac{280\text{ mL/L}}{1000\text{ mL/L} - 280\text{ mL/L}} = 0.38$$

13.186 (7,520 mg/L) (x MGD RAS) (8.34) = (2,200 mg/L) (6.4 MGD + x MGD RAS) (8.34)
(7,520) (x MGD) = (2,200) (6.4 + x MGD)
$7{,}520x = 14{,}080 + 2{,}200x$
$5{,}320x = 14{,}080$
$x = 2.65$ MGD

13.187 $$\frac{3400 \text{ lb/day BOD}}{(\text{x lb/day MLSS})\ (69/100)} = 0.5$$
$x = 9{,}565$ lb MLSS desired

13.188 (2,710 mg/L) (0.79 MG) (8.34 lb/gal) = 17,855 lb MLSS
17,855 lb MLSS – 14,900 lb MLSS = 2,955 lb MLSS to be wasted

13.189 $$\frac{(110 \text{ mg/L})\ (3.10 \text{ MGD})\ (8.34 \text{ lb/gal})}{(\text{x lbs MLSS})\ (68/100)} = 0.4$$
$x = 10{,}456$ lb MLSS desired
(2,200 mg/L) (1.1 MG) (8.34 lb/gal) = 20,183 lb MLSS actual
20,183 lb MLSS actual – 10,456 lb MLSS desired = 9,727 lb MLSS to be wasted

13.190 $$\frac{\text{x lb MLSS}}{3220 \text{ lb/day}} = 5.6 \text{ days desired sludge age}$$
$x = 18{,}032$ lb MLSS desired
(2,900 mg/L) (0.910 MG) (8.34 lb/gal) + 22,009 lb MLSS actual
22,009 lb MLSS actual – 18,032 lb MLSS desired = 3,977 lb MLSS to be wasted

13.191 $$\frac{32{,}400 \text{ lb MLSS}}{\text{x lb/day WAS} + (23 \text{ mg/L})\ (3.22 \text{ MGD})\ (8.34 \text{ lb/gal})} = 9 \text{ days}$$
$$\text{WAS Pumping Rate} = \frac{32{,}400 \text{ lb}}{\text{x lb/day} + 618 \text{ lb/day}} 9 \text{ days}$$
$$\frac{32{,}400}{9} = \text{x} + 618$$
$3{,}600 = x + 618$
$2{,}982 \text{ lb/day} = x$

13.192 (6,640 mg/L) (x MGD) (8.34 lb/gal) = 5,580 lb/day dry solids
$x = 0.10$ MGD

13.193 (6,200 mg/L) (x MGD) (8.34 lb/gal) = 8,710 lb/day dry solids
$x = 0.17$ MGD

13.194 $$\frac{(2725 \text{ mg/L})\ (1.8 \text{ MG})\ (8.34 \text{ lb/gal})}{(7420 \text{ mg/L})\ (\text{x MGD})\ (8.34 \text{ lb/gal}) + (18 \text{ mg/L})\ (4.3 \text{ MGD})\ (8.34 \text{ lb/gal})} = 9 \text{days}$$
$$\frac{40{,}908 \text{ lb MLSS}}{(7420 \text{ mg/L})\ (\text{x MGD})\ (8.34 \text{ lb/gal}) + 646 \text{ lb/day}} = 9 \text{days}$$
$$\frac{40{,}908}{9} = (7420)\ (\text{x})\ (8.34) + 646$$

$4,545 = (7,420)\ (x)\ (8.34) + 646$
$3,899 = (7,420)\ (x)\ (8.34)$
$x = 0.063\ \text{MGD}$

13.195 $$\frac{(2610\ \text{mg/L})\ (1.7\ \text{MG})\ (8.34\ \text{lb/gal})}{(6140\ \text{mg/L})\ (\text{x MGD})\ (8.34\ \text{lb/gal}) + (14\ \text{mg/L})\ (3.8\ \text{MGD})\ (8.34\ \text{lb/gal})} = 8.5\ \text{days}$$

$$\frac{37{,}005\ \text{lb MLSS}}{(6140\ \text{mg/L})\ (\text{x MGD})\ (8.34\ \text{lb/gal}) + (444\ \text{lb/day})} = 8.5\ \text{days}$$

$$\frac{37{,}005}{8.5} = (6140)\ (\text{x})\ (8.34) + (444)$$

$4,354 = (6,140)\ (x)\ (8.34) + (444)$
$3,910 = (6,140)\ (x)\ (8.34)$
$x = 0.076\ \text{MGD}$

13.196 $$\frac{166{,}000\ \text{gal}}{7917\ \text{gph}} = 21\ \text{hrs}$$

13.197 $$\frac{370{,}000\ \text{gal}}{9583\ \text{gph}} = 39\ \text{hrs}$$

13.198 $$\frac{420{,}000\ \text{gal}}{12{,}708\ \text{gph}} = 33\ \text{hrs}$$

13.199 $$\frac{210{,}000\ \text{gal}}{12{,}917\ \text{gph}} = 16\ \text{hrs}$$

13.200 (80 ft) (4 ft) (15 ft) (7.48 gal/cu ft) = 359,040 gal
13.201 (220 mg/L) (1.72 MGD) (8.34 lb/gal) = 3,156 lb/day

13.202 $$\frac{(222\ \text{mg/L})\ (0.399\ \text{MGD})\ (8.34\ \text{lb/gal})}{(3340)\ (0.22\ \text{MG})\ (8.34\ \text{lb/gal})\ (68/100)} = 0.18$$

13.203 (0.785) (90 ft) (90 ft) (12 ft) (7.48 gal/cu ft) = 570, 739 gal
13.204 (160 mg/L) (3.92 MGD) (8.34 lb/gal) = 5,231 lb/day

13.205 $$\frac{(2700\ \text{mg/L})\ (0.53\ \text{MG})\ (8.34\ \text{lb/gal})}{(190\ \text{mg/L})\ (1.8\ \text{MGD})\ (8.34\ \text{lb/day})} = 4.2\ \text{days}$$

13.206 $$\frac{410\ \text{mL}}{2100\ \text{mL} - 440\ \text{mL}} = \frac{440\ \text{mL}}{1660\ \text{mL}} = 0.265$$

(0.265) (6.1 MGD) = 1.62 MGD

$$\frac{1.62\ \text{MG}}{1} \times \frac{1\ \text{day}}{1440\ \text{mL}} \times \frac{1{,}000{,}000\ \text{gal}}{1\ \text{million gal}} = 1125\ \text{gpm}$$

13.207 $$\begin{array}{ll} \text{MLSS} & 2100\ \text{mg/L} \\ & \underline{-2050\ \text{mg/L}} \\ & \quad 50\ \text{mg/L(in excess)} \end{array}$$

(50 mg/L) (0.45 MG) (8.34 lb/gal) = 188 lb
188 lb/day = (4,920 mg/L) (additional WAS, MGD) (8.34 lb/gal)

$$\frac{188}{(4920)\ (8.34)} = 0.0046\ \text{MGD (additional WAS)}$$

new WAS = 0.120 MGD + 0.0046 MGD = 0.125 MGD

13.208 (-80) we need an extra 80 mg/L MLSS in aeration.
(80 mg/L) (0.44 MG) (8.34 lb/gal) = 294 lb (needed)
294 lb/day = (4,870 mg/L) (excess WAS, MGD) (8.34 lb/gal)

$$\frac{294}{(4870)\ (8.34)} = 0.007\ \text{MGD}\ (\text{Excess WAS})$$

$$\text{WAS} = \frac{(87.3\ \text{gpm})\ (1440\ \text{min/day})}{1{,}000{,}000\ \text{gal}}$$

= 0.126 MGD
new WAS = 0.126 MGD – 0.007 MGD = 0.119 MGD

$$\frac{(0.119\ \text{MGD})\ (1{,}000{,}000\ \text{gal})}{1440\ \text{min}} = 83\ \text{gpm}$$

13.209

$$\frac{(2210\ \text{mg/L})\ (0.66\ \text{MG})\ (8.34\ \text{lb/gal})}{(131\ \text{mg/L})\ (3.25\ \text{MGD})\ (8.34\ \text{lb/gal})}$$

$$\text{Sludge Age} = \frac{1459\ \text{lb}}{426\ \text{lb/day}} = 3.42\ \text{days}$$

13.210

$$\frac{(146\ \text{mg/L})\ (2.88\ \text{MGD})\ (8.34\ \text{lb/gal})}{\text{x lb MLVSS}} = 0.6$$

x = 5,845 lb MLVSS

13.211

$$\frac{310{,}000\ \text{gal}}{17{,}083\ \text{gph}} = 18\ \text{hr}$$

13.212

$$\frac{(161\ \text{mg/L})\ (2.41\ \text{MGD})\ (8.34\ \text{lb/gal})}{\text{x lb MLVSS}} = 0.8$$

x = 4,045 lb MLVSS

13.213

$$\frac{(2910\ \text{mg/L})\ (0.46\ \text{MGD})\ (8.34\ \text{lb/gal})}{(170\ \text{mg/L})\ (1.4\ \text{MG})\ (8.34\ \text{lb/gal})} = 5.8\ \text{days}$$

13.214

$$\frac{620{,}000\ \text{gal}}{15{,}000\ \text{gph}} = 41.3\ \text{lb}$$

13.215

$$\frac{(3980\ \text{mg/L})\ (0.26\ \text{MGD})\ (8.34\ \text{lb/gal})}{(200\ \text{mg/L})\ (0.4\ \text{MGD})\ (8.34\ \text{lb/gal})} = 13.0\ \text{days}$$

13.216 (2,710 mg/L) (0.44 MG) (8.34 lb/gal) = 9,945 lb SS

13.217

$$\frac{(146\ \text{mg/L})\ (2.88\ \text{MGD})\ (8.34\ \text{lb/gal})}{\text{x lb MLVSS}} = 0.4$$

x = 8,767 lb MLVSS

13.218 (2,510 mg/L) (0.59 MG) (8.34 lb/gal) = 12,351 lb MLVSS

13.219

$$\frac{(2740\ \text{mg/L})\ (0.710\ \text{MGD})\ (8.34\ \text{lb/gal})}{(184\ \text{mg/L})\ (1.86\ \text{MGD})\ (8.34\ \text{lb/gal})} = 5.7\ \text{days}$$

13.220 $\frac{(2680\text{ mg/L})\ (1.41\text{ MG})\ (8.34\text{ lb/gal}) + (1910\text{ mg/L})\ (0.118\text{ MG})\ (8.34\text{ lb/gal})}{(5870\text{ mg/L})\ (0.076\text{ MGD})\ (8.34\text{ lb/gal}) + (20\text{ mg/L})\ (3.1\text{ MGD})\ (8.34\text{ lb/gal})}$

$= \frac{31{,}515\text{ MLSS} + 1854\text{ lb MLSS}}{3721\text{ lb/day SS} + 517\text{ lb/day SS}} = 7.9\text{ days}$

13.221 $\frac{231\text{ mL/L}}{769\text{ mL/L}} = 0.30$

13.222 $\frac{3720\text{ lb/day}}{(x\text{ lb MLSS})\ (70/100)} = 0.5$

$x = 10{,}629$ lb MLSS

13.223 $\frac{x\text{ lb MLSS}}{3740\text{ lb/day SS}} = 5\text{ days}$

$x = 18{,}700$ lb MLSS

(2,810 mg/L) (0.78 MG) (8.34 lb/gal) = 18,280 lb MLSS

No MLSS should be wasted.

13.224 (6,410 mg/L) (x MGD) (8.34 lb/gal) = 4,110 lb/day

$x = 0.077$ MGD

13.225 (250 mg/L) (0.41 MGD) (8.34 lb/gal) = 855 lb/day

13.226 (161 mg/L) (0.225 MGD) (8.34 lb/gal) = 302 lb/day

13.227 (223 mg/L) (0.259 MGD) (8.34 lb/gal) = 482 lb/day

13.228 (200 mg/L) (0.19 MGD) (8.34 lb/gal) = 317 lb/day

13.229 $\frac{(192\text{ mg/L})\ (0.219\text{ MGD})\ (8.34\text{ lb/gal})}{7.8\text{ ac}} = 46\text{ lb/day/ac}$

13.230 $\frac{(145\text{ mg/L})\ (0.167\text{ MGD})\ (8.34\text{ lb/gal})}{7.1\text{ ac}} = 28.4\text{ lb/day/ac}$

13.231 $\frac{(128\text{ mg/L})\ (0.072\text{ gpd})\ (8.34\text{ lb/gal})}{2.2\text{ ac}} = 35\text{ lb/day/ac}$

13.232 $\frac{(189\text{ mg/L})\ (x\text{ MGD})\ (8.34\text{ lb/gal})}{15\text{ ac}} = 22\text{ lb}$

$x = 0.21$ MGD

13.233 $\frac{169\text{ mg/L Removed}}{210\text{ mg/L Total}} \times 100 = 80\%\text{ BOD Removed}$

13.234 $\frac{140\text{ mg/L}}{267\text{ mg/L}} \times 100 = 52\%$

13.235 $\frac{246\text{ mg/L}}{290\text{ mg/L}} \times 100 = 85\%\text{ BOD Removed}$

13.236 $\frac{84\text{ mg/L Removed}}{142\text{ mg/L Total}} \times 100 = 59\%\text{ BOD Removed}$

13.237 $\frac{3.6\text{ ac-ft/day}}{22\text{ ac}} = 0.164\text{ ft/day}$

= (0.165 ft/day) (12 in/ft) = 2 in/day

13.238 $\frac{6 \text{ ac-ft/day}}{16 \text{ ac}} = 0.38 \text{ ft/day}$

$(0.38 \text{ ft/day})(12 \text{ in/ft}) = 4.6 \text{ in/day}$

13.239 $\frac{2{,}410{,}000 \text{ gpd}}{(7.48 \text{ gal/cu ft})(43{,}560 \text{ cu ft/ac-ft})} = 7.4 \text{ ac-ft}$

$\frac{7.4 \text{ ac-ft/day}}{17 \text{ ac}} = 0.44 \text{ ft/day}$

$(0.44 \text{ ft/day})(12 \text{ in/ft}) = 5.3 \text{ in/day}$

13.240 $(16 \text{ ac})(43{,}560 \text{ sq ft/ac}) = 696{,}960 \text{ sq ft}$

$\frac{1{,}880{,}000 \text{ gpd}}{696{,}960 \text{ sq ft}} = 2.70 \text{ gpd/sq ft}$

$\frac{(2.70 \text{ gpd/sq ft})(1.6 \text{ in./day})}{\text{gpd/sq ft}} = 4.3 \text{ in./day}$

13.241 $\frac{1340 \text{ people}}{5 \text{ ac}} = 268 \text{ people/ac}$

13.242 $\frac{5580 \text{ people}}{19 \text{ ac}} = 294 \text{ people/ac}$

13.243 $\frac{(1640 \text{ mg/L})(0.8 \text{ MGD})(8.34 \text{ lb/gal})}{0.2 \text{ lb BOD/day/person}} = 54{,}710 \text{ people}$

13.244 $\frac{(2260 \text{ mg/L})(0.257 \text{ MGD})(8.34 \text{ lb/gal})}{0.2 \text{ lb BOD/day/person}} = 24{,}220 \text{ people}$

13.245 $\frac{19 \text{ ac-ft}}{0.44 \text{ ac-ft/day}} = 43 \text{ days}$

13.246 $\frac{(450 \text{ ft})(700 \text{ ft})(8 \text{ ft})(7.48 \text{ gal/cu ft})}{300{,}000 \text{ gpd}} = 63 \text{ days}$

13.247 $\frac{(250 \text{ ft})(400 \text{ ft})(6 \text{ ft})(7.48 \text{ gal/cu ft})}{72{,}000 \text{ gpd}} = 62 \text{ days}$

13.248 $\frac{33 \text{ ac}}{0.48 \text{ ac-ft/day}} = 68 \text{ days}$

13.249 $(720 \text{ ft})(460 \text{ ft})(6 \text{ ft})(7.48 \text{ gal/cu ft}) = 14{,}864{,}256 \text{ gal}$

$\frac{14{,}864{,}256 \text{ gal}}{310{,}000 \text{ gal/day}} = 48 \text{ days}$

13.250 $(705 \text{ ft})(430 \text{ ft})(4.17 \text{ ft}) = 1{,}264{,}136 \text{ cu ft}$

$(0.50 \text{ ac-ft/day})(43{,}560 \text{ cu ft}) = 21{,}780 \text{ cu ft/day}$

$\frac{1{,}264{,}136 \text{ cu ft}}{21{,}780 \text{ cu ft}} = 58 \text{ days}$

13.251 $(0.16 \text{ MGD})(171 \text{ mg/L})(8.34 \text{ lb/gal}) = 228.2 \text{ lb/day}$

$$\frac{(698\text{ ft})\ (395\text{ ft})}{43{,}560\text{ sq ft}} = 6.33\text{ ac}$$

$$\frac{228.2\text{ lb/day}}{6.33\text{ ac}} = 36.1\text{ lb/day ac}$$

13.252 $$\frac{(750\text{ ft})\ (435\text{ ft})}{43{,}560\text{ sq ft}} = 7.49\text{ ac}$$

$$\frac{(0.79\text{ ac-ft/day})\ (12\text{ in./ft})}{7.49\text{ ac}} = 1.27\text{ in./day}$$

13.253 (192 mg/L) (0.37 MGD) (8.34 lb/day) = 592 lb/day

13.254 $$\frac{(240\text{ mg/L})\ (0.285\text{ MGD})\ (8.34\text{ lb/gal})}{9.1\text{ ac}} = 63\text{ lb/day/ac}$$

13.255 $$\frac{176\text{ mg/L BOD Removed}}{220\text{ mg/L BOD Total}} \times 100 = 80\%\text{ BOD Removed}$$

13.256 $$\frac{3.8\text{ ac-ft/day}}{22\text{ ac}} = 0.17\text{ ft/day}$$

(0.17 ft/day) (12 in/day) = 2 in/day

13.257 $$\frac{93\text{ mg/L Removed}}{166\text{ mg/L Total}} \times 100 = 56\%$$

13.258 (222 mg/L) (0.302 MGD) (8.34 lb/gal) = 559 lb/day

13.259 $$\frac{(135\text{ mg/L})\ (0.080\text{ MGD})\ (8.34\text{ lb/gal})}{\frac{(400\text{ ft})\ (220\text{ ft})}{43{,}560\text{ sq ft/ac}}} = 45\text{ lb/day/ac}$$

13.260 $$\frac{1{,}980{,}000\text{ gpd}}{(21\text{ ac})\ (43{,}560\text{ sq ft/ac})} = 2.2\text{ gpd/sq ft}$$

$$\frac{(2.2\text{ gpd/sq ft})\ (1.6\text{ in./day})}{\text{gpd/sq ft}} = 3.5\text{ in./day}$$

13.261 $$\frac{6200\text{ people}}{22\text{ ac}} = 282\text{ people/ac}$$

13.262 $$\frac{18.4\text{ ac-ft}}{0.52\text{ ac-ft/day}} = 35\text{ days}$$

13.263 $$\frac{(2910\text{ mg/L})\ (0.9\text{ MGD})\ (8.34\text{ lb/gal})}{0.4\text{ lb BOD/day/person}} = 54{,}606\text{ people}$$

13.264 $$\frac{(440\text{ ft})\ (730\text{ ft})\ (6\text{ ft})\ (7.48\text{ gal/cu ft})}{450{,}000\text{ gpd}} = 32\text{ days}$$

13.265 (3.4 mg/L) (4.6 MGD) (8.34 lb/gal) = 130 lb/day

13.266 (11 mg/L) (1.68 MGD) (8.34 lb/gal) = 154 lb/day

13.267 (2,200 mg/L) (0.200 MGD) (8.34 lb/gal) = 3,578 lb/day

13.268 $(x \text{ mg/L})\ (5.12 \text{ MGD})\ (8.34 \text{ lb/gal}) = 320 \text{ lb/day}\ x = 7.5 \text{ mg/L}$

13.269 4.9 mg/L + 0.8 mg/L = 5.7 mg/L

13.270 $8.8 \text{ mg/L} = x \text{ mg/L} + 0.9 \text{ mg/L}$

$x = 7.9 \text{ mg/L}$

13.271 7.9 mg/L + 0.6 mg/L = 8.5 mg/L

13.272 (10.7 mg/L) (4.0 MGD) (8.34 lb/gal) = 357 lb/day

13.273 $$\frac{(11.1 \text{ mg/L})\ (2.88 \text{ MGD})\ (8.34 \text{ lb/gal})}{(65/100)} = 410 \text{ lb/day}$$

13.274 $$\frac{(9.8 \text{ mg/L})\ (4.1 \text{ MGD})\ (8.34 \text{ lb/gal})}{(60/100)} = 559 \text{ lb/day}$$

13.275 $$\frac{(19 \text{ mg/L})\ (1.724 \text{ MGD})\ (8.34 \text{ lb/gal})}{(65/100)} = 420 \text{ lb/day}$$

13.276 $$\frac{(\text{x mg/L})\ (5.65 \text{ MGD})\ (8.34 \text{ lb/gal})}{(65/100)} = 950 \text{ lb}$$

$x = 13 \text{ mg/L}$

13.277 $$\frac{0.75 \text{ lb}}{(16 \text{ gal})\ (8.34 \text{ lb/gal}) + (0.75 \text{ lb})} \times 100 = 0.56\%$$

13.278 $$\frac{\text{x lb}}{(24 \text{ gal})\ (8.34 \text{ lb/gal}) + \text{x lb}} \times 100 = 0.9$$

$$\frac{100 \text{ x}}{200 \text{ lb} + \text{x lb}} = 0.9$$

$100x = (0.9)\ (200 + x)$

$100x = 180 + 0.9x$

$99.1x = 180$

$x = 1.8 \text{ lb}$

13.279 160 grams = 0.35 lb

$$\frac{0.35 \text{ lb}}{(12 \text{ gal})\ (8.34 \text{ lb/gal}) + 0.3 \text{ lb}} \times 100 = 0.3\%$$

13.280 $(10/100)\ (x \text{ lb liquid polymer}) = (0.5/100)\ (172 \text{ lb polymer solution})$

$x = 8.6 \text{ lb liquid polymer}$

13.281 $(10/100)\ (x \text{ gal liquid polymer})\ (10.4 \text{ lb/gal}) = (0.3/100)\ (55 \text{ gal poly sol})\ (8.34 \text{ lb/gal}) = 1.3 \text{ gal liquid polymer}$

13.282 $$\frac{(10/100)\ (26 \text{ lb}) + (0.5/100)\ (110 \text{ lb})}{26 \text{ lb} + 110 \text{ lb}} \times 100$$

$$\frac{2.28 \text{ lb}}{136 \text{ lb}} \times 100 = 2.3\% \text{ Strength}$$

13.283 $$\frac{(12/100)\ (6 \text{ gal})\ (10.2 \text{ lb/gal}) + (0.3/100)\ (30 \text{ gal})\ (8.4 \text{ lb/gal})}{(6 \text{ gal})\ (10.2 \text{ lb/gal}) + (30 \text{ gal})\ (8.34 \text{ lb/gal})} \times 100$$

$$\frac{7.3 \text{ lb} + 0.8 \text{ lb}}{61 \text{ lb} + 250 \text{ lb}} \times 100 = 2.6\% \text{ Strength}$$

13.284 $$\frac{(10/100)\ (12\ \text{gal})\ (10.2\ \text{lb/gal}) + (0.28/100)\ (42\ \text{gal})\ (8.34\ \text{lb/gal})}{(12\ \text{gal})\ (10.2\ \text{lb/gal}) + (42\ \text{gal})\ (8.34\ \text{lb/gal})} \times 100$$

$$= \frac{12.24\ \text{lb} + 0.98\ \text{lb}}{122.4\ \text{lb} + 350.3\ \text{lb}} \times 100 = 2.8\%\ \text{Strength}$$

13.285 (10 mg/L) (4.10 MGD) (8.34 lb/gal) = 342 lb/day alum required

$$\frac{342\ \text{lb/day}}{5.88\ \text{lb alum/gal solution}} = 58\ \text{gpd solution}$$

13.286 (8 mg/L) (1.44 MGD) (8.34 lb/gal) = 96 lb/gal alum required

$$\frac{96\ \text{lb/day}}{6.15\ \text{lb}} = 15.6\ \text{gpd solution}$$

13.287 (11 mg/L) (2.13 MGD) (8.34 lb/gal) = (600,000 mg/L) (x MGD) (8.34 lb/gal)

$x = 0.000039$ MGD or 39 gpd solution

13.288 (9 mg/L) (4.44 MGD) (8.34 lb/gal) = (600,000 mg/L) (x MGD) (8.34 lb/gal)

$x = 0.0000666$ MGD or 66.6 gpd

13.289 $$\frac{30\ \text{gpm}}{80\ \text{gpm}} \times 100 = 37.5\%$$

13.290 $$\frac{22\ \text{gpm}}{80\ \text{gpm}} \times 100 = 27.5\%$$

13.291 $$\frac{14\ \text{gpm}}{70\ \text{gpm}} \times 100 = 20\%$$

13.292 $$\frac{40\ \text{gpm}}{110\ \text{gpm}} \times 100 = 36\%$$

13.293 $$\frac{(35\ \text{gpd})\ (3785\ \text{mL/gal})}{1440\ \text{min/day}} = 92\ \text{mL/min}$$

13.294 $$\frac{(45\ \text{gal/day})\ (3785\ \text{mL/gal})}{1440\ \text{min/day}} = 118\ \text{mL/min}$$

13.295 (9 mg/L) (0.91 MGD) (8.34 lb/gal) = (600,000 mg/L) (x MGD) (8.34 lb/gal)

$x = 0.000136$ MGD or 13.6 gpd

$$\frac{(13.6\ \text{gal/day})\ (3785\ \text{mL/gal})}{1440\ \text{min/day}} = 36\ \text{mL/min}$$

13.296 (11 mg/L) (1.42 MGD) (8.34 lb/gal) = (600,000 mg/L) (x MGD) (8.34 lb/gal)

$x = 0.000026$ MGD or 26 gpd

$$\frac{(26\ \text{gal/day})\ (3785\ \text{mL/gal})}{1440\ \text{min/day}} = 68\ \text{mL/min}$$

13.297 $$\frac{2.1\ \text{lb}}{30\ \text{min}} = 0.07\ \text{lb/min}$$

(0.07 lb/min) (1,440 min/day) = 101 lb/day

13.298 $\frac{1.5\text{ lb}}{30\text{ min}} = 0.05\text{ lb/min}$

(0.05 lb/min) (1,440 min/day) = 72 lb/day

13.299 12 oz = 0.75 lb

$$\begin{array}{l} 2.10\text{ lb container} + \text{chemical} \\ \underline{-.75\text{ lb container}} \\ 1.35\text{ lb chemical} \end{array}$$

$$\frac{1.35\text{ lb chemical}}{30\text{ min}} = 0.045\text{ lb/min}$$

(0.045 lb/min) (1,440 min/day) = 65 lb/day

13.300

$$\begin{array}{l} 2.5\text{ lb chemical} + \text{container} \\ \underline{-0.5\text{ lb container}} \\ 2.0\text{ lb chemical} \end{array}$$

$$\frac{2.0\text{ lb chemical}}{30\text{ min}} = 0.067\text{ lb/min}$$

(0.067 lb/min) (1,440 min/day) = 96.5 lb/day

13.301 $\frac{770\text{ mL}}{5\text{ min}} = 154\text{ mL/min}$

$$\frac{(154\text{ mL/min})\ (1\text{ gal})\ (1440\text{ min/day})}{3785\text{ mL}} = 59\text{ gpd Solution}$$

(14,000 mg/L) (0.000059 MGD) (8.34 lb/gal) = 6.9 lb/day polymer

13.302 $\frac{900\text{ mL}}{5\text{ min}} = 180\text{ mL/min}$

$$\frac{(180\text{ mL})}{\text{minute}} \times \frac{(1\text{L})}{1000\text{ mL}} \times \frac{(1\text{ gal})}{3785\text{ L}} \times (1440\text{ min/day}) = 66.0\text{ gpd Solution}$$

(12,000 mg/L) (0.000066 MGD) (8.34 lb/gal) = 6.6 lb/day

13.303 $\frac{610\text{ mL}}{5\text{ min}} = 122\text{ mL/min}$

$$\frac{(120\text{ mL/min})\ (1\text{ gal})}{3785\text{ L}}(1440\text{ min/day}) = 46\text{ gpd Solution}$$

(13,000 mg/L) (0.000046 MGD) (8.34 lb/gal) (1.2 specific gravity) = 6.0 lb/day

13.304 $\frac{800\text{ mL}}{5\text{ min}} = 160\text{ mL/min}$

$$\frac{(160\text{ mL})}{\text{min}} \times \frac{(1\text{ gal})}{3785\text{ L}} \times \frac{(1440\text{ min})}{\text{day}} = 61\text{ gpd Solution}$$

(5,000 mg/L) (0.000061 MGD) (8.34 lb/gal) (1.15 specific gravity) = 2.9 lb/day

13.305 $$\frac{(0.785)\ (4\ \text{ft})\ (4\ \text{ft})\ (1.5\ \text{ft})\ (7.48\ \text{gal/cu ft})}{3\ \text{min}} = 47\ \text{gpm}$$

13.306 $$\frac{(0.785)\ (5\ \text{ft})\ (5\ \text{ft})\ (1.25\ \text{ft})\ (7.48\ \text{gal/cu ft})}{5\ \text{min}} = 37\ \text{gpm}$$

13.307 $$\frac{(0.785)\ (15\ \text{ft})\ (5\ \text{ft})\ (\text{x ft})\ (7.48\ \text{gal/cu ft})}{4\ \text{min}} = 30$$

13.308 $$\frac{(0.785)\ (5\ \text{ft})\ (5\ \text{ft})\ (1.6\ \text{ft})\ (7.48\ \text{gal/cu ft})}{3\ \text{min}} = 78\ \text{gpm}$$

13.309 $$\frac{537}{7\ \text{day}} = 77\ \text{lb/day average}$$

13.310 $$\frac{2300\ \text{lb}}{115\ \text{lb/day}} = 20\ \text{days}$$

13.311 $$\frac{1002\ \text{lb}}{66\ \text{lb/day}} = 15.2\ \text{days}$$

13.312 (0.785) (5 ft) (5 ft) (3.4 ft) (7.48 gal/cu ft) = 499 gal

$$\frac{499\ \text{gal}}{97\ \text{gpd}} = 5.1\ \text{days}$$

13.313 (11 mg/L) (3.75 MGD) (8.34 lb/gal) = 344 lb/day

13.314 $$\frac{(7.1\ \text{mg/L})\ (3.24\ \text{MGD})\ (8.34\ \text{lb/gal})}{0.65} = 295\ \text{lb/day}$$

13.315 $$\frac{\text{x lb}}{(32\ \text{gal})\ (8.34\ \text{lb/gal}) + \text{x lbs}} \times 100 = 0.2$$

$$\frac{100\text{x}}{267\ \text{lb} + \text{x}} = 0.2$$

$100x = (0.2)\ (267\ \text{lb} + x)$

$100x = 53.4 + 0.2x$

$100x - 0.2x = 53.4$

$99.8x = 53.4$

$x = 53.4\ \text{lb}$

13.316 $$\frac{1.9\ \text{lb}}{30\ \text{min}} = 0.063\ \text{lb/min}$$

(0.063 lb/min) (1,440 min/day) = 90.7 lb/day

13.317 (12 mg/L) (2.75 MGD) (8.34 lb/gal) = 275 lb/day alum

$$\frac{275\ \text{lb/day}}{5.88\ \text{lb alum/gal solution}} = 47\ \text{gpd solution}$$

13.318 (x mg/L) (5.115 MGD) (8.34 lb/gal) = 379 lb

$x = 8.8\ \text{mg/L}$

13.319 12 oz = 0.75 lb; 2 lb 6 oz = 2.38 lb

2.38 lb chemical + container

-.75 lb container

1.63 lb chemical

$$\frac{1.63\text{ lb}}{30\text{ min}} \times 0.054\text{ lb/min}$$

(0.054 lb/min) (1,440 min/day) = 78 lb/day

13.320 (10 mg/L) (3.244 MGD) (8.34 lb/gal) = (600,000 mg/L) (x MGD) (8.34 lb/gal)

x = 0.000054 MGD, or 54 gpd

13.321 $$\frac{32\text{ gpm}}{90\text{ gpm}} \times 100 = 35.5\%$$

13.322 7.8 mg/L = x mg/L + 0.5 mg/L7.3 mg/L = x

13.323 (12/100) (x gal) (9.6 lb/gal) = (0.4/100) (60 gal) (8.34 lb/gal)

x = 1.7 gal

13.324 $$\frac{660\text{ mL}}{5\text{ min}} = 132\text{ mL/min}$$

$$\frac{(132\text{ mL/min})\ (1\text{ gal})\ (1440\text{ min/day})}{3785\text{ L}} = 50.2\text{ gpd}$$

(12,000 mg/L) (0.0000502 MGD) (8.34 lb/gal) = 5 lb/day polymer

13.325 $$\frac{(0.785)\ (6\text{ ft})\ (\text{x ft})\ (7.48\text{ gal/cu ft})}{5\text{ min}} = 30\text{ gpm}$$

x = 0.71 ft

13.326 $$\frac{20\text{ gpm}}{90\text{ gpm}} \times 100 = 22\%$$

13.327 (9.6 mg/L) (4.3 MGD) (8.34 lb/gal) = 344 lb/day

13.328 $$\frac{2100\text{ lb}}{90\text{ lb/day}} = 23.3\text{ days}$$

13.329 $$\frac{(50\text{ gal/day})\ (3785\text{ mL/gal})\ (1\text{ day})}{1440\text{ min}} = 131\text{ mL/min}$$

13.330 $$\frac{888\text{ mL}}{5\text{ min}} = 178\text{ mL/min}$$

$$\frac{(178\text{ mL/min})\ (1\text{ gal})\ (1440\text{ min/day})}{3785\text{ L}} = 68\text{ gpd}$$

(9,000 mg/L) (0.000068 MGD) (8.34 lb/gal) = 5.1 lb/day polymer

13.331 (9 mg/L) (3.22 MGD) (8.34 lb/gal) = (600,000 mg/L) (x MGD) (8.34 lb/gal)

x = 0.0000483 MGD or 48 gpd

$$\frac{(48\text{ gal/gal})\ (3785\text{ mL/gal})\ (1\text{ day})}{1440\text{ min}} = 126\text{ mL/min}$$

13.332 $\dfrac{(0.785)\ (4\text{ ft})\ (4\text{ ft})\ (1.25\text{ ft})\ (7.48\text{ gal/cu ft})}{3\text{ min}} = 39\text{ gpm}$

13.333 $\dfrac{(11.1\text{ mg/L})\ (3.115\text{ MGD})\ (8.34\text{ lb/gal})}{0.65} = 444\text{ lb/day hypochlorite}$

13.334 $\dfrac{(12/100)\ (6\text{ gal})\ (11.2\text{ lb/gal}) + (0.3/100)\ (22\text{ gal})\ (8.34\text{ lb/gal})}{(6\text{ gal})\ (11.2\text{ lb/gal}) + (22\text{ gal})\ (8.34\text{ lb/gal})} \times 100$

$\dfrac{8.1\text{ lb} + 0.55\text{ lb}}{67.2 + 183.5} \times 100 = 3.45\%\text{ Strength}$

13.335 (160 mg/L) (4.82 MGD) (8.34 lb/gal) = 6,432 lb/day solids

13.336 (184 mg/L rem.) (3.9 MGD) (8.34 lb/gal) = 5,985 lb/day solids

13.337 (135 mg/L BOD rem.) (2.1 MGD) (8.34 lb/gal) = 2,364 lb/day BOD removed

$\dfrac{0.5\text{ lb/day SS}}{1\text{ lb/day BOD Removal}} = \dfrac{x\text{ lb/day SS}}{2364\text{ lb/day BOD Removed}}$

$x = (0.5)\ (2{,}364)$

$x = 1{,}182$ lb/day dry solids

13.338 (157 mg/L BOD rem.) (2.84 MGD) (8.34 lb/gal) = 3,496 lb/day BOD rem.

Use the Y-value:

$\dfrac{0.66\text{ lb/day SS}}{1\text{ lb/day BOD Rem}} = \dfrac{x\text{ lb/day SS}}{3496\text{ lb/day BOD Removed}}$

$(0.66)\ (3{,}496) = x$

$x = 2{,}307$ lb/day dry solids

13.339 $x = \dfrac{0.71\text{ grams}}{31\text{ grams}} \times 100$

$x = 2.3\%$

13.340 $4.1 = \dfrac{8{,}520\text{ lb/day solids}}{x\text{ lb/day sludge}} \times 100$

$x = 202{,}857$ lb/day sludge

13.341 $5.5 = \dfrac{x\text{ lb/day solids}}{(9350\text{-gal sludge})\ (8.34\text{ lb/gal}} \times 100$

$x = 4{,}289$ lb/day solids

13.342 $5.3 = \dfrac{1490\text{ lb/day solids}}{(x\text{ gpd sludge})\ (8.34\text{ lb/gal})} \times 100$

1,490 lb/day solids

$x = 3{,}371$ gpd sludge

13.343 $4.4 = \dfrac{900\text{ lb/day solids}}{(x\text{ gpd sludge})\ (8.34\text{ lb/gal}} \times 100$

$x = 2{,}452$ gpd sludge

13.344 (20,100 lb/day) (0.41) = (x lb/day) (0.06)

$$\frac{(20{,}100)\ (0.04)}{0.06} = \mathrm{x}$$

x = 13,400 lb/day thickened sludge

13.345 (2,910 gpd) (8.34 lb/gal) (0.051) = (x gpd) (8.64 lb/gal) (0.06)

$$\frac{(2910)\ (8.34)\ (0.051)}{(8.64)\ (0.06)} = \mathrm{x}$$

x = 2,388 gpd thickened sludge

13.346 (12,400 lb/day) (0.034) = (x lb/day) (0.054)

$$\frac{(12{,}400)\ (0.034)}{0.054} = \mathrm{x}$$

x = 7,807 thickened sludge

13.347 (6,100 gpd) (8.34 lb/gal) (0.041) = (x gpd) (8.6 lb/gal) (0.064)

$$\frac{(6100)\ (8.34)\ (0.041)}{(8.6)\ (0.064)} = \mathrm{x}$$

x = 3,793 gpd thickened sludge

13.348

$$\frac{(70 \text{ gpm} + 82 \text{ gpm})\ (1440 \text{ min/day})}{(0.785)\ (28 \text{ ft})\ (28 \text{ ft})}$$

$$= \frac{(152 \text{ gpm})\ (1440 \text{ min/day})}{615 \text{ sq ft}}$$

= 356 gpd/sq ft

13.349

$$\frac{(162 \text{ gpm})\ (1440 \text{ min/day})}{(0.785)\ (28 \text{ ft})\ (28 \text{ ft})} = 379 \text{ gpd/sq ft}$$

13.350

$$\frac{(122{,}000 \text{ gpd})\ (8.34 \text{ lb/gal})\ (4.1/100)}{(0.785)\ (44 \text{ ft})\ (44 \text{ ft})} = 27 \text{ lb/day/sq ft}$$

13.351

$$\frac{(60 \text{ gpm})\ (1440 \text{ min/day})\ (8.34 \text{ lb/gal})\ (3.8/100)}{(0.785)\ (32 \text{ ft})\ (32 \text{ ft})} = 34 \text{ lb/day/sq ft}$$

13.352

$$\frac{(0.785)\ (46 \text{ ft})\ (46 \text{ ft})\ (3.8 \text{ ft})\ (7.48 \text{ gal/cu ft})}{(28 \text{ gpm})\ (1440 \text{ min/day})} = 1.2 \text{ days}$$

13.353

$$\frac{(0.785)\ (40 \text{ ft})\ (40 \text{ ft})\ (4.3 \text{ ft}\ (7.48 \text{ gal/cu ft})}{(31 \text{ gpm})\ (60 \text{ min/hr})} = 21.7$$

13.354 4% = 40,000 mg/L

0.9% = 9,000 mg/L

$$\frac{31{,}000 \text{ mg/L Rem.}}{40{,}000 \text{ mg/L}} \times 100 = 78\%$$

13.355 $\frac{8.4}{3.3} = 2.5$

13.356 $\frac{8.0}{3.1} = 2.6$

13.357 Solids in: (130 gpm) (1,440 min/day) (8.34 lb/gal) (3.6/100) = 56,205 lb/day
Solids out: (50 gpm) (1,440 min/day) (8.34 lb/gal) (8.1/100) = 48,639 lb/day
Sol. out of thick: (130 gpm – 50 gpm) (1,440 min/day) (8.34) (0.059/100) = 567 lb/day
Solids entering = 56,205 lb/day
Solids leaving = 48,639 lb day + 567 lb/day = 49,206 lb/day (Because more solids are entering than are leaving, the sludge blanket will increase.)

13.358 (a) Solids in: (110 gpm) (1,440 min/day) (8.34 lb/gal) (3.6/100) = 47,558 lb/day
Underflow solids out: (65 gpm) (1,440 min/day) (8.34 lb/gal) (7.1/100) = 55,424 lb/day
Eff. Sol. Out: (110 gpm – 65 gpm) (1,440 min/day) (8.34) (0.052/100) = 281 lb/day
Solids entering = 47,558 lb/day
Solids out = (55,424 lb/day + 281 lb/day) 55,705 lb/day (Because more solids are leaving than are entering, the blanket will decrease.)
(b) The pounds per day decrease is:

55,705 lb/day
-47,558 lb/day
8,147 lb/day

13.359 $\frac{9400 \text{ lb/day}}{24 \text{ hr/day}} = 392 \text{ lb/hr}$

$$\text{Fill time, hr} = \frac{(0.785)\ (30\text{ ft})\ (30\text{ ft})\ (1.8\text{ ft})\ (7.48\text{ gal/cu ft})\ (8.34)\ (6.6/100)}{392\text{ lb/hr}} = 13.4\text{ hr}$$

13.360 $\frac{14{,}000 \text{ lb/day}}{24 \text{ hr/day}} = 583 \text{ lb/hr}$

$$\text{Fill time, hr} = \frac{(0.785)\ (30\,ft)\ (30\,ft)\ (2.5\,ft)\ (7.48\,gal/cu\,ft)\ (8.34)\ (8/100}{583\text{ lb/hr}} = 15.1\text{ hr}$$

13.361 $\frac{2.6 \text{ ft}}{6 \text{ ft}} = \frac{\text{x lb/min}}{60 \text{ lb/min}}$

$\frac{(2.6)\ (60)}{6} = \text{x}$

x = 26 lb/min solids storage rate
60 lb/min = x lb/min withdrawal + 28 lb/min storage
x = 32 lb/min solids withdrawal
(x gpm) (8.34 lb/gal) (5.6/100) = 32 lb/min

$$x = \frac{32}{(8.34)\ (0.056)}$$

x = 69 gpm sludge qithdrawal

13.362 $\frac{3.3\text{ ft}}{7\text{ ft}} = \frac{\text{x lb/min}}{61\text{ lb/min}}$

$$\frac{(3.3)\ (61)}{7} = x$$

29 lb/min = x solids storage rate

61 lb/min = x lb/min withdrawal + 29 lb/min storage

29 lb/min = x solids withdrawal

(x gpm) (8.34 lb/gal) (5.6/100) = 29 lb/min

$$x = \frac{29}{(8.34)\ (0.056)}$$

x = 63 gpm sludge withdrawal

13.363 $\frac{910\text{ gpm}}{(0.785)\ (40\text{ ft})\ (40\text{ ft})} = 0.7\text{ gpm/sq ft}$

13.364 $\frac{660\text{ gpm}}{(0.785)\ (30\text{ ft})\ (30\text{ ft})} = 0.9\text{ gpm/sq ft}$

13.365 $\frac{(8420\text{ mg/L})\ (0.17\text{ MGD})\ (8.34\text{ lb/gal})}{(0.785)\ (40\text{ ft})\ (40\text{ ft})} = 9.5\text{ lb/day/sq ft}$

$$\frac{9.5\text{ lb/day/sq ft}}{24\text{ hr/day}} = 0.40\text{ lb/hr/sq ft}$$

13.366 $\frac{(120\text{ gpm})\ (60\text{ min/hr})\ (8.34\text{ lb/gal})\ (0.7/100)}{(65\text{ ft})\ (20\text{ ft})} = 0.3\text{ lb/hr/sq ft}$

13.367 (9 cfm) (60 min/hr) (0.075 lb/hr) = 41 lb/hr

13.368 (12 cfm) (60 min/hr) (0.075 lb/cu ft) = 54 lb/hr

13.369 8,600 mg/L = 0.86% solids

$$\text{Air-to-Solids Ratio} = \frac{(8\text{ cfm})\ (0.075\text{ lb/cu ft})}{(85\text{ gpm})\ (8.34\text{ lb/gal})\ (0.86/100)} = 0.10$$

13.370 7,800 mg/L = 0.78% solids

$$\text{Air-to-Solids Ratio} = \frac{(5\text{ cfm})\ (0.075\text{ lb/cu ft})}{(60\text{ gpm})\ (8.34\text{ lb/gal})\ (0.78/100)} = 0.10$$

13.371 $\frac{90\text{ gpm}}{85\text{ gpm}} \times 100 = 106\%$

13.372 $112 = \frac{\text{x gpm}}{70\text{ gpm}} \times 100$

$$\frac{(112)\ (70)}{100} = x$$

x = 78 gpm

13.373 $\frac{7460 \text{ mg/L Solids Removed}}{7700 \text{ mg/L in Influent}} \times 100 = 97\%$

13.374 $\frac{4.8}{0.841} = 5.6$

13.375 (40 gpm) (60 min/hr) = 2,400 gph

13.376 $\frac{86{,}400 \text{ gpd}}{24 \text{ hr/day}} = 3600 \text{ gph}$

13.377 (70gpm) (60 min/hr) (30 min/31 min) = 4,065 gph

13.378 $\frac{78{,}000 \text{ gpd} \times (25 \text{ min})}{24 \text{ hr/day} \times (27 \text{ min})} = 2990 \text{ gph}$

13.379 7,600 mg/L = 0.76%

$$\frac{(110{,}000 \text{ gpd})}{24 \text{ hr/day}} = (8.34 \text{ lb/gal}) \ (0.76/100) = 291 \text{ lb/hr}$$

13.380 $(80 \text{ gpm}) \ (60 \text{ min/hr}) \ (8.34 \text{ lb/gal}) \ (0.75/100) \frac{(30 \text{ min})}{(32 \text{ min})} = 281 \text{ lb/hr}$

13.381 $\frac{(32 \text{ cu ft}) \ (7.48 \text{ gal/cu ft}) \ (8.34 \text{ lb/gal}) \ (6.6/100)}{(70 \text{ gpm}) \ (8.34 \text{ lbs/gal}) \ (0.73/100)} = 31 \text{ min}$

13.382 $\frac{(22 \text{ cu ft}) \ (7.48 \text{ gal/cu ft}) \ (8.34 \text{ lbs/gal}) \ (9/100)}{(55 \text{ gpm}) \ (8.34 \text{ lb/gal}) \ (0.76/100)} = 31 \text{ min}$

13.383 $\frac{7200 \text{ mg/L}}{8000 \text{ mg/L}} \times 100 = 90\%$

13.384 $\frac{(16 \text{ cu ft}) \ (62.4 \text{ lb/cu ft}) \ (4.4/100) + (4.0 \text{ cu ft}) \ (62.4 \text{ lb/cu ft}) \ (8.0/100)}{(16 \text{ cu ft}) \ (62.4 \text{ lb}) + (4.0 \text{ cu ft}) \ (62.4 \text{ lb/cu ft})} \times 100$

$$\frac{43.9 \text{ lb} + 19.9 \text{ lb}}{998 \text{ lb} + 250 \text{ lb}} \times 100 = 5.1\%$$

13.385 $\frac{(12 \text{ cu ft}) \ (62.4 \text{ lb/cu ft}) \ (3.8/100) + (4 \text{ cu ft}) \ (62.4 \text{ lb/cu ft}) \ (8.0/100)}{(12 \text{ cu ft}) \ (62.4 \text{ lb/cu ft}) + (4 \text{ cu ft}) \ (62.4 \text{ lb lbs/cu ft})} \times 100$

$$\frac{28.4 \text{ lb} + 19.9 \text{ lb}}{748.8 \text{ lb} + 249.6 \text{ lb}} \times 100 = 4.8\%$$

13.386 (48,400 gal/d) (8.34 lb/gal) (0.8/100) = 3,229 lb/d

13.387 (0.785) (24 ft) (24 ft) = 452 sq ft

$$\frac{(170 \text{ gpm}) \ (1440 \text{ min/day})}{452 \text{ sq ft}} = 542 \text{ gpd/sq ft}$$

13.388 (240 gpm) (1,440 min/d) (8.34 lb/gal) (1.3/100) = 37,470 lb/d

(0.785) (40 ft) (40 ft) = 1,256 sq ft

$$\frac{37{,}470 \text{ lb/d}}{1256 \text{ sq ft}} = 30 \text{ lb/d/sq ft}$$

13.389 $\dfrac{690\ \text{gpm}}{(0.785)\ (34\ \text{ft})\ (34\ \text{ft})} = 0.76\ \text{gpm/sq ft}$

13.390 (130 gal/min) (60 min/hr) (8.34 lb/gal) (0.98/100) = 637.5 lb/hr

(0.785) (30 ft) (30 ft) = 706.5 sq ft

$\dfrac{637.5}{706.5} = 0.90\ \text{lb/hr/sq ft}$

13.391 (181 mg/L removed) (3.5 MGD) (8.34 lb/gal) = 5,283 lb/day SS

13.392 $\dfrac{0.66\ \text{grams}}{32\ \text{grams}} \times 100 = 2.1\%$

13.393 (3,750 gpd) (8.34 lb/gal) (3.9/100) = (x gpd) (8.34 lb/gal) (8/100)

$\dfrac{(3750)\ (8.34)\ (0.039)}{(8.34)\ (0.08)} = \text{x}$

x = 1,828 gpd flow

13.394 (9,550 gal) (8.34 lb/gal) (4.9/100) = 3,903 lb/d solids

13.395 (132 mg/L BOD rem.) (2.96 MGD) (8.34 lb/gal) = 3,259 lb/day BOD rem.

$\dfrac{0.5\ \text{lb SS}}{1\ \text{lb BOD Rem}} = \dfrac{\text{x lb/d SS}}{3259\ \text{lb/d BOD Rem.}}$

(0.5) (3,259) = x

x = 1,630 lb/d SS produced

13.396 $\dfrac{(0.785)\ (42\ \text{ft})\ (42\ \text{ft})\ (5\ \text{ft})\ (7.48\ \text{gal/cu ft})}{(32\ \text{gpm})\ (60\ \text{min/hr})} = 27\ \text{hr}$

13.397 $\dfrac{7.7}{3.1} = 2.5$

$\dfrac{(7920\ \text{mg/L})\ (0.14\ \text{MGD})\ (8.34\ \text{lb/gal})}{(0.785)\ (36\ \text{ft})\ (36\ \text{ft})} = 9.1\ \text{lb/d/sq ft}$

13.398 $\dfrac{6780\ \text{mg/L}}{7010\ \text{mg/L}} \times 100 = 97\%$

13.399 $\dfrac{(170\ \text{gpm})\ (1440\ \text{min/d})}{(0.785)\ (30\ \text{ft})\ (30\ \text{ft})} = 326\ \text{gpd/sq ft}$

13.400 $\dfrac{3.0}{3.3} \times 100 = 91\%$

13.401 (9 cfm) (60 min/hr) (0.075 lb/cu ft) = 41 lb/hr

13.402 (110 gpm) (1,440 min/d) (8.34 lb/gal) (4/100) = 52,842 lb/d

From underflow: (50 gpm) (1,440 min/day) (8.34 lb/gal) (7.7/100) = 46,237 lb/d

From eff. flow: (110 gpm – 50 gpm) (1,440 min/d) (8.34) (0.070/100) = 504 lb/d

In: 52,842 lb/d

Out: 46,237 + 504 = 46,741 lb/d

(Because more solids are entering the thickener than leaving, the sludge blanket will increase.)

13.403 $\frac{(60 \text{ gpm})(1440 \text{ min/d})(8.34 \text{ lb})(4.1/100)}{(0.785)(32 \text{ ft})(32 \text{ ft})} = 36.7 \text{ lb/d/sq ft}$

13.404 $\frac{190{,}000 \text{ gpd}}{1440 \text{ min/d}} = 132 \text{ gpmt}$

$\frac{132 \text{ gpm}}{(60 \text{ ft})(14 \text{ ft})} = 0.16 \text{ gpm/sq ft}$

13.405 (0.785) (26 ft) (26 ft) (2.6 ft) (7.48 gal/cu ft) (8.34 lb/gal) (6.9/100)

$\frac{9400 \text{ lb/d}}{24 \text{ hr/d}} = 15.2 \text{ hrs}$

13.406 $\frac{84{,}000 \text{ gpd}}{24 \text{ hr/d}} = 3500 \text{ gph}$

13.407 $\frac{(6 \text{ cfm})(0.075 \text{ lb/cu ft})}{(110 \text{ gpm})(8.34 \text{ lb/gal})(0.81/100)} = 0.06$

13.408 $112 = \frac{\text{x gpm}}{74 \text{ gpm}} \times 100$

$\frac{(112)(74)}{100} = \text{x}$

$x = 83$ gpm

13.409 $\frac{(79{,}000 \text{ gpd})}{24 \text{ hr/d}} \times \frac{(32 \text{ min})}{34 \text{ min}} = 3090 \text{ gal/hr}$

13.410 $\frac{2.5 \text{ ft}}{6 \text{ ft}} = \frac{\text{x lb/min stored solids}}{48 \text{ lb/min solids entering}}$

$\frac{(2.5)(48)}{6} = \text{x}$

$x = 20$ lb/min storage rate
48 lb/min = x lb/min + 20 lb/min
$x = 28$ lb/min solids withdrawal
Sludge withdrawal: (x gpm) (8.34 lb/gal) (8/100) = 20 lb/min solids withdrawal
$x = 41$ gpm sludge withdrawal rate

13.411 $\frac{(110{,}000 \text{ gpd})(8.34 \text{ lb/gal})(0.711/100)}{24 \text{ hr/d}}$

= 272 lb/hr

13.412 $\frac{(34 \text{ cu ft})(7.48 \text{ gal/cu ft})(8.34 \text{ lb/gal})(6.6/100)}{(70 \text{ gpm})(8.34 \text{ lb/gal})(0.73/100)} = 32.5 \text{ min}$

13.413 $$\frac{(100\text{ gpm})\ (60\text{ min/hr})\ (8.34\text{ lb/gal})\ (0.79/100)\ (24\text{ min})}{(25.5\text{ min})} = 372\text{ lb/hr}$$

13.414 $$\frac{(12\text{ cu ft})\ (62.4\text{ lb/cu ft})\ (3.9/100)\ +\ (5\text{ cu ft})\ (62.4\text{ lb/cu ft})\ (7.8/100)}{(12\text{ cu ft})\ (62.4\text{ lb/cu ft})\ +\ (5\text{ cu ft})\ (62.4\text{ lb/cu ft})} \times 100$$

$$\frac{29.2\text{ lb} + 24.3\text{ lb}}{749\text{ lb} + 312\text{ lb}} \times 100 = 5\%$$

13.415 $$\frac{(4240\text{ gpd})\ (8.34\text{ lb/gal})\ (5.9/100) + (6810\text{ gpd})\ (8.34\text{ lb/gal})\ (3.5/100)}{(4120\text{ gpd})\ (8.34\text{ lb/gal}) + (6810\text{ yd})\ (8.34\text{ lb/gal})} \times 100$$

$$\frac{2086\text{ lb/d} + 1988\text{ lb/d}}{34{,}361\text{ lb/d} + 56{,}795\text{ lb/d}} \times 100 = 4.5\%$$

13.416 $$\frac{(3510\text{ gpd})\ (8.34\text{ lb/gal})\ (5.2/100) + (5210\text{ gpd})\ (8.34\text{ lb/gal})\ (4.1/100)}{(3510\text{ gpd})\ (8.34\text{ lb/gal}) + (5210\text{ gpd})\ (8.34\text{ lb/gal})} \times 100$$

$$= \frac{1517\text{ lb/d} + 1782\text{ lb/d}}{29{,}273\text{ lb/d} + 43{,}451\text{ lb/d}} \times 100 = 4.5\%$$

13.417 $$\frac{(3910\text{ gpd})\ (8.34\text{ lb/gal})\ (6.3/100) + (6690\text{ gpd})\ (8.34\text{ lb/gal})\ (4.9/100)}{(3910\text{ gpd})\ (8.35\text{ lb/gal}) + (6690\text{ gpd})\ (8.34\text{ lb/gal})} \times 100$$

$$= \frac{(2054\text{ lb/d}) + (2734\text{ lb/d})}{(32{,}609\text{ lb/d}) + (55{,}795\text{ lb/d})} \times 100 = 5.4\%$$

13.418 $$\frac{(2510\text{ gpd})\ (8.34\text{ lb/gal})\ (4.3/100) + (3600\text{ gpd})\ (8.60\text{ lb/gal})\ (6.1/100)}{(2510\text{ gpd})\ (8.34\text{ lb/gal}) + (3600\text{ gpd})\ (8.60\text{ lb/gal})} \times 100$$

$$= \frac{900\text{ lb/d} + 1889\text{ lb/d}}{20{,}993\text{ lb/d} + (30{,}960\text{ lb/d})} \times 100 = 5.4\%$$

13.419 (0.9 gal/stroke) (30 strokes/min) = 27 gpm

13.420 (0.785) (0.83 ft) (0.83 ft) (0.25 ft) (7.48 gal/cu ft) = 1 gal/stroke
(1 gal/stroke) (30 strokes/min) = 30 gpm

13.421 (0.785) (0.67 ft) (0.67 ft) (0.25 ft) (7.48 gal/cu ft) = 0.7 gal/stroke
(0.7 gal/stroke) (32 strokes/min) (120 min/day) = 2,688 gpd

13.422 (0.785) (1 ft) (1 ft) (0.33 ft) (7.48 gal/cu ft) = 1.9 gal/stroke
(91.9 gal/stroke) (32 strokes/min) (140 min/day) = 8,512 gpd

13.423 $$\frac{(130\text{ mg/L SS Rem.})\ (2.5\text{ MGD})\ (8.34\text{ lb/gal})}{0.035} = (132\text{ gpm})\ (\text{x min/d})\ (8.34)$$

$$\frac{(130)\ (2.5)\ (8.34)}{(0.035)\ (32)\ (8.34)} = \text{x}$$

$x = 290$ min/d

$$\frac{290\text{ min/d}}{24\text{ hr/d}} = 12\text{ min/hr}$$

13.424 $\frac{(120\text{ mg/L SS})(1.87\text{ MGD})(8.34\text{ lb/gal})}{0.036} = (28\text{ gpm})(x\text{ min/d})(8.34\text{ lb/d})$

$$\frac{(120)(1.87)(8.34)}{(0.038)(28)(8.34)} = 204\text{ min/d}$$

$$\frac{204\text{ min/d}}{24\text{ hrs/d}} = 8.5\text{ min/hr}$$

13.425 $\frac{(124\text{ mg/L Rem.})(3.48\text{ MGD})(8.34\text{ lb/gal})}{(4.0/100)} = (38\text{ gpm})(x\text{ min/d})(8.34\text{ lb/d})$

$$x = \frac{(124)(3.48)(8.34)}{(0.04)(38)(8.34)} = 284\text{ min/d}$$

$$\frac{284\text{ min/d}}{24\text{ hrs/d}} = 11.8\text{ min/hr}$$

13.426 (8,620 lb/d solids) (66/100) = 5,689 lb/d VS
13.427 (2,810 lb/d solids) (67/100) = 1,883 lb/d VS
13.428 (3,720 gpd sludge) (8.34 lb/gal) (5.8/100) (70/100) = 1,260 lb/d VS
13.429 (5,115 gpd sludge) (8.34 lb/gal) (7/100) (67/100) = 2,001 lb/d

13.430 $25 = \frac{x\text{ gal}}{295{,}200\text{ gal}} \times 100$

295,200 gal x = (295,200 yd) (25/100) x = 73,800 gal

13.431 $21 = \frac{x\text{ gal}}{(0.785)(40\text{ ft})(40\text{ ft})(24\text{ ft})(7.48\text{ gal/cu ft})} \times 100$

x = (0.785) (40 ft) (40 ft) (24 ft) (7.48 gal/cu ft) (21/100) x = 47,350 gal

13.432 $x = \frac{62{,}200\text{ gal}}{(0.785)(50\text{ ft})(50\text{ ft})(20\text{ ft})(7.48\text{ gal/cu ft})} \times 100$

x = 21%

13.433 $20 = \frac{x\text{ gal}}{(0.785)(40\text{ ft})(40\text{ ft})(18\text{ ft})(7.48\text{ gal/cu ft})} \times 100$

x = (0.785) (40 ft) (40 ft) (18 ft) (7.48 gal/cu ft) (20/100)
x = 33,822 gal

13.434 $\frac{(66{,}130\text{ lb/day Sludge})(5.3/100)(70/100)}{(120{,}000\text{ gal})(8.34\text{ lb})(6.3/100)(56/100)}$

$$= \frac{2379\text{ lb VS/day}}{35{,}308\text{ lb VS}}$$

= 0.07 lb VS added/day/lb VS in digester

13.435 $0.06\text{ lb/d/lb VS} = \frac{x\text{ lb/d VS}}{(22{,}310\text{ gal})(8.34\text{ lb/gal})(6.2/100)(55/100)}$

(0.06) (22,130) (8.34 lb/gal) (6.2/100) (55/100) = x lb/day
x = 381 lb VS/d

13.436 $$\frac{(60{,}400\text{ lb/d Sludge})\ (5.4/100)\ (67/100)}{(96{,}000\text{ gal})\ (8.34\text{ lb/gal})\ (5/100)\ (58/100)}$$

$$= \frac{2185\text{ lb VS/d}}{23{,}219\text{ lb VS}}$$

$$= 0.09\text{ lb VS zdded/d/lb VS in digester}$$

13.437 $$\frac{0.07\text{ lb VS Added/d}}{\text{lb VS in Digester}} = \frac{(900\text{ gpd})\ (8.34\text{ lb/gal})\ (5.5/100)\ (69/100)}{(x\text{ gal})\ (8.34\text{ lb/gal})\ (8.2/100)\ (52/100)}$$

$$\text{x} = \frac{(900)\ (8.34)\ (0.055)\ (0.69)}{(0.07)\ (8.80)\ (0.082)\ (0.52)}$$

x = 10,962-gal seed sludge

13.438 $$\frac{(86{,}100\text{ lb/day Sludge})\ (5/100)\ (7/100)}{(0.785)\ (50\text{ ft})\ (50\text{ ft})\ (22\text{ ft})}$$

$$= 0.07\text{ lb VS/day cu ft}$$

13.439 $$\frac{(28{,}500\text{ gpd})\ (8.34\text{ lb/gal})\ (5.5/100)\ (72/100)}{(0.785)\ (40\text{ ft})\ (40\text{ ft})\ (22\text{ ft})}$$

$$= \frac{0.347\text{ lb VS/day}}{\text{cu ft}} \times 1000$$

$$= \frac{(0.347\text{ lb VS/day})}{1\text{ cu ft}1000}$$

$$= \frac{347\text{ lb VS/day}}{1000\text{ cu ft}}$$

13.440 $$\frac{(36{,}220\text{ gpd})\ (8.34\text{ lb/gal})\ (5.6/100)\ (68/100)}{(0.785)\ (50\text{ ft})\ (50\text{ ft})\ (20\text{ ft})}$$

$$\frac{0.293\text{ lb VS/d}}{1\text{ cu ft}} \times 1000$$

$$\frac{293\text{ lb/d VS}}{1000\text{ cu ft}}$$

13.441 $$\frac{(16{,}200\text{ gpd Sludge})\ (8.34\text{ lb/gal})\ (5.1/100)\ (72/100)}{(0.785)\ (50\text{ ft})\ (50\text{ ft})\ (18\text{ ft})}$$

$$= \frac{0.14\text{ lb/d VS}}{\text{cu ft}}$$

$$= \frac{(0.14\text{ lb/d VS})}{1\text{ cu ft}} \times 1000$$

$$= \frac{140\text{ lb/d VS}}{1000\text{ cu ft}}$$

13.442 $\dfrac{116/\text{d VS}}{10 \text{ lb Digester Sludge}} = \dfrac{(2600 \text{ gpd})\ (8.34 \text{ lb/gal})\ (5.7/100)\ (66/100)}{\text{x lb Digested Sludge}}$

$x = (2{,}600 \text{ gpd})\ (8.34)\ (0.057)\ (0.66)\ (10)$

$x = 8{,}158$ lb digested sludge

13.443 $\dfrac{1 \text{ lb/d VS}}{10 \text{ lb Digester Sludge}} = \dfrac{(6300 \text{ gpd})\ (8.34 \text{ lb/gal})\ (5/100)\ (70/100)}{\text{x lb Digested Solids}}$

$x = (6{,}300)\ (8.34)\ (0.05)\ (0.70)\ (10)$

$x = 18{,}390$ lb digested sludge

13.444 $\dfrac{1 \text{ lb/day VS}}{10 \text{ lb Digester Sludge}} = \dfrac{(5200 \text{ gpd})\ (8.34 \text{ lb/gal})\ (6.5/100)\ (67/100)}{\text{x lb Digested Sludge}}$

$x = (5{,}200)\ (8.34)\ (0.065)\ (0.67)\ (10)$

$x = 18{,}887$ lb digested sludge

13.445 $\dfrac{1 \text{ lb/d VS}}{10 \text{ lb Digester Sludge}} = \dfrac{(3800 \text{ gpd})\ (8.34 \text{ lb/gal})\ (6/100)\ (72/100)}{\text{x lb Digested Sludge}}$

$x = (3{,}800)\ (8.34)\ (0.06)\ (0.72)\ (10)$

$x = 13{,}691$ lb digested sludge

13.446 $\dfrac{174 \text{ mg/L}}{2220 \text{ mg/L}} = 0.08$

13.447 $\dfrac{160 \text{ mg/L}}{2510 \text{ mg/L}} = 0.06$

13.448 $\dfrac{144 \text{ mg/L}}{2410 \text{ mg/L}} = 0.06$

13.449 $\dfrac{178 \text{ mg/L}}{2620 \text{ mg/L}} = 0.07$

13.450 (2,280 mg/L) (0.244 MG) (8.34 lb/gal) = 4,640 lb lime

13.451 (2,010 mg/L) (0.200 MG) (8.34 lb/gal) = 3,353 lb lime

13.452 (2,540 mg/L) (0.234 MG) (8.34 lb/gal) = 4,898 lb lime

13.453 (2,410 mg/L) (0.182 MG) (8.34 lb/gal) = 3,658 lb lime

13.454 $\dfrac{0.68 - 0.52}{0.68 - (0.68)\ (0.52)} \times 100$

$\dfrac{0.16}{0.3264} \times 100 = 49\%$ VS Reduction

13.455 $\dfrac{0.70 - 054}{0.70 - (0.70)\ (0.54)} \times 100$

$\dfrac{0.16}{0.322} \times 100 = 50\%$ VS Reduction

13.456 $\dfrac{0.70-0.53}{0.70-(0.70)\ (0.53)}\times 100$

$\dfrac{0.17}{0.329}\times 100 = 52\%$ VS Reduction

13.457 $\dfrac{0.69-0.54}{0.69-(0.69)\ (0.54)}\times 100$

$\dfrac{0.15}{0.3174}\times 100 = 47\%$

13.458 $\dfrac{(3800\text{ gpd})\ (8.34\text{ lb/gal})\ (6.3/100)\ (73/100)\ (57/100)}{36{,}500\text{ cu ft}}$

$= 0.02$ lb VS Destroyed/d/cu ft

13.459 $\dfrac{(4520\text{ gpd})\ (8.34\text{ lb/gal})\ (7/100)\ (69/100)\ (54/100)}{33{,}000\text{ cu ft}}$

$= 0.03$ lb VS Destroyed/d/cu ft

13.460 $\dfrac{(2600\text{ gpd})\ (8.34\text{ lb/gal})\ (5.6/100)\ (72/100)\ (52/100)}{(0.785)\ (50\text{ ft})\ (50\text{ ft})\ (18\text{ ft})}$

$= 0.01$ lb VS Destroyed/d/cu ft

13.461 $\dfrac{(2800\text{ gpd})\ (8.34\text{ lb/gal})\ (6.1/100)\ (65/100)\ (56/100)}{(0.785)\ (40\text{ ft})\ (40\text{ ft})\ (17\text{ ft})}$

$= 0.024$ lb VS Destroyed/d/cu ft

$\dfrac{0.024\text{ lb VS Destroyed/day}}{1\text{ cu ft}}\times 1000 = \dfrac{24\text{ lb VS Destroyed}}{1000\text{ cu ft}}$

13.462 $\dfrac{6600\text{ cu ft gas/d}}{500\text{ lb VS destroyed/d}} = 13.2$ cu ft/lb VS destroyed

13.463 $\dfrac{19{,}330\text{ cu ft gas/day}}{2110\text{ lb VS/d}\ (59/100)} = 15.5$ cu ft/lb VS Destroyed

13.464 $\dfrac{8710\text{ cu ft gas/d}}{582\text{ lb/d VS Destroyed/d}} = 15$ cu ft lb VS Destroyed

13.465 $\dfrac{26{,}100\text{ cu ft gas/d}}{(3320\text{ lb VS/d})\ (54/100)} = 14.6$ cu ft/lb VS Destroyed

13.466 $\dfrac{(0.785)\ (40\text{ ft})\ (40\text{ ft})\ (12\text{ ft})\ (7.48\text{ gal/cu ft})}{9100\text{ gpd}} = 12.4\text{ d}$

13.467 $\dfrac{(0.785)\ (40\text{ ft})\ (40\text{ ft})\ (10\text{ ft})\ (7.48\text{ gal/cu ft})}{8250\text{ gpd}} = 11.4\text{ d}$

13.468 $\dfrac{(80\text{ ft})\ (25\text{ ft})\ (12\text{ ft})\ (7.48\text{ gal/cu ft})}{7800\text{ gpd}} = 23\text{ d}$

13.469 $3.4\%\ Solids: \dfrac{(0.785)\ (30\text{ ft})\ (30\text{ ft})\ (12\text{ ft})\ (7.48\text{ gal/cu ft})}{11{,}000\text{ gpd}} = 5.8\text{ d}$

$6\%\ Solids: \dfrac{(0.785)\ (30\text{ ft})\ (30\text{ ft})\ (12\text{ ft})\ (7.48\text{ gal/cu ft})}{5400\text{ gpd}} = 11.7\text{ d}$

Higher solids content permits greater time for digestion.

13.470 $\dfrac{40\text{ cfm}}{1000\text{ cu ft}} = \dfrac{\text{x cfm Air Removed}}{(0.785)\ (70\text{ ft})\ (70\text{ ft})\ (10\text{ ft})}$

$\text{x} = \dfrac{(40)\ (0.785)\ (70\text{ ft})\ (70\text{ ft})\ (10\text{ ft})}{1000} = 1539\text{ cfm Air}$

13.471 $\text{Oxygen Uptake, mg/L/hr} = \dfrac{5.4\text{ mg/L} - 3.4\text{ mg/L}}{3\text{ min}} \times 60\text{ min/hr}$

$\dfrac{(2)\ (60)}{3} = 40\text{ mg/L/hr}$

13.472 $\text{Oxygen Uptake, mg/L/hr} = \dfrac{5.7\text{ mg/L} - 3.6\text{ mg/L}}{3\text{ min}} \times 60\text{ min/hr}$

$\dfrac{(2.1)\ (60)}{3} = 42\text{ mg/L/hr}$

13.473 (22 mg/L) (0.106 MG) (8.34 lb/gal) = 19.4 lb Caustic

13.474 (16 mg/L) (0.148 MG) (8.34 lb/gal) = 19.7 lb Caustic

13.475 $\dfrac{(64\text{ mg})}{2\text{ L}}(0.054\text{ MG})\ (8.34\text{ lb/gal}) = 14.4\text{ lb Caustic}$

13.476 (0.785) (60 ft) (60 ft) (14 ft) (7.48 gal/cu ft) = 295,939 gal

$\dfrac{(90\text{ mg})}{2\text{ L}}(0.296\text{ MG})\ (8.34\text{ lb/gal}) = 111\text{ lb Caustic}$

13.477 (3.6 gpm) (1,440 min/d) (8.34 lb/gal) (5.1/100) (71/100) = 1,566 lb VS/day

13.478 (0.785) (55 ft) (55 ft) (22 ft) = 52,242 cu ft (47,200 gal/d) (8.34 lb/d) (5.3/100) (71/100) = 14,813 lb VS/d

$\dfrac{14{,}813\text{ lb VS/d}}{52{,}242\text{ cu ft}} = 0.28\text{ lb VS cu ft/d}$

13.479 $\dfrac{181\text{ mg/L}}{2120\text{ mg/L}} = 0.085$

13.480 756,000 L = 0.756 mL

0.756 mL × 1,820 mg/L = 1,376 kg

13.481 $\%\text{ VS Reduction} = \dfrac{0.67 - 0.55}{0.67 - (0.67 \times 0.55)} \times 100\%$

= 39.8%

13.482 (2,600 kg VS) (100/9.5) (100/66) (1L/1.14 kg) =

$$\frac{(2600)\ (10.5)\ (1.5)}{1.14} = 35{,}921\ \text{L}$$

13.483
$$\frac{1\ \text{lb/d VS}}{10\ \text{lb Digested Sludge}} = \frac{(8200\ \text{gpd})\ (8.34\ \text{lb/gal})\ (5.7/100)\ (65/100)}{\text{x lb Digested Sludge}}$$

$x = (8{,}200)\ (8.34)\ (0.057)\ (0.65)\ (10)$

$x = 25{,}338$ lb

13.484 (4,400 lb/d Solids) (67/100) = 2,948 lb/d VS

13.485
$$\frac{(12{,}900\ \text{gpd})\ (8.34\ \text{lb/gal})\ (5.4/100)\ (65/100)}{\dfrac{(0.785)\ (60\ \text{ft})\ (60\ \text{ft})\ (20\ \text{ft})}{1000}} = \frac{67\ \text{lb/d VS}}{1000\text{-cu ft}}$$

13.486
$$\frac{(4040\ \text{gpd})\ (8.34\ \text{lb/gal})\ (5.4/100) + (5820\ \text{gpd})\ (8.34\ \text{lb/gal})\ (3.3/100)}{(4040\ \text{gpd})\ (8.34\ \text{lb/gal}) + (5820\ \text{gpd})\ (8.34\ \text{lb/gal})} \times 100$$

$$\frac{1819\ \text{lb/d} + 1602\ \text{lb/d}}{33{,}694\ \text{lb/d} + 45{,}339\ \text{lb/d}} \times 100 = 4.3\%$$

13.487 (0.785) (0.67 ft) (0.67 ft) (0.5 ft) (7.48 gal/cu ft) = 1.3 gal/stroke

(1.3 gal/stroke) (3,500 strokes/d) = 4,550 gpd

13.488
$$\text{x} = \frac{88{,}200\ \text{gal}}{(0.785)\ (60\ \text{ft})\ (60\ \text{ft})\ (24\ \text{ft})\ (7.48\ \text{gal/cu ft})} \times 100$$

$x = 17.4\%$

13.489
$$\frac{(3800\ \text{gpd sludge})\ (8.34\ \text{lb/gal})\ (4.1/100)\ (70/100)\ (54/100)}{36{,}000\ \text{cu ft}}$$

= 0.01 lb/d VS destroyed/cu ft

13.490
$$\frac{156\ \text{mg/L}}{2310\ \text{mg/L}} = 0.07$$

13.491 (2,240 mg/L) (0.24 MG) (8.34 lb/gal) = 4,484 lb lime

13.492
$$24 = \frac{\text{x gal}}{(0.785)\ (50\ \text{ft})\ (50\ \text{ft})\ (22\ \text{ft})\ (7.48\ \text{gal/cu ft})} \times 100$$

$$= \frac{(24)\ (0.785)\ (50)\ (50)\ (22)\ (7.48)}{100}$$

$x = 77{,}508$ gal

13.493 (4,310 gpd sludge) (8.34 lb/gal) (5.3/100) (72/100) = 1,372 lb/d VS

13.494
$$\frac{1447\ \text{lb/day} + 1496\ \text{lb/d}}{24{,}520\ \text{lb/d} + 39{,}365\ \text{lb/d}} \times 100 = 4.6\%$$

13.495
$$\frac{150\ \text{mg/L}}{2470\ \text{mg/L}} = 0.06$$

13.496
$$\frac{(42{,}500\ \text{lb/d Sludge})\ (4/100)\ (60/100)}{(8.34)\ (6/100)\ (55/100)} =$$

0.04 lb VS Added/d lb VS in Digester (94,000 gal)

13.497 (0.785) (0.75 ft) (0.75 ft) (0.42 ft) (7.48 gal/cu ft) = 1.4 gal/stroke
(1.4 gal/stroke) (30 strokes/min) = 42 gpm

13.498 $$\frac{(19{,}200 \text{ gpd})\ (8.34 \text{ lb/gal})\ (5/100)\ (66/100)}{\frac{(0.785)\ (40 \text{ ft})\ (40 \text{ ft})\ (21 \text{ ft})}{1000}} = \frac{200 \text{ lb VS Added/d}}{1000 \text{ cu ft}}$$

13.499 (2,200 mg/L) (0.3 MG) (8.34 lb/gal) = 5,504 lb Lime

13.500 $$\frac{6760 \text{ cu ft}}{580 \text{ lb VS Destroyed/d}} = 11.7 \text{ cu ft Gas lb VS Destroyed}$$

13.501 $$\frac{0.67 - 0.52}{0.67 - (0.67)\ (0.52)} \times 100 = 47\%$$

13.502 $$\frac{0.09 \text{ lb/day VS Added}}{1 \text{ lb VS in Digester}} = \frac{(1230 \text{ gpd})\ (8.34 \text{ lb/gal})\ (4.1/100)\ (66/100)}{(\text{x gal})\ (8.5 \text{ lb/gal})\ (7.5/100)\ (55/100)}$$

$$x = \frac{(1230)\ (8.34)\ (0.041)\ (0.66)}{(0.09)\ (8.5)\ (0.075)\ (0.55)}$$

x = 8688 gal

13.503 $$\frac{0.70 - 0.56}{0.70 - (0.70)\ (0.56)} \times 100 = 45\%$$

13.504 $$\frac{(0.785)\ (60 \text{ ft})\ (60 \text{ ft})\ (12 \text{ ft})\ (7.48 \text{ gal/cu ft})}{9350 \text{ gpd}} = 27.1 \text{ d}$$

13.505 $$\frac{22{,}400 \text{ cu ft gal/d}}{(2610 \text{ lb VS/d})\ (56/100)} = 15 \text{ cu ft/lb Destroyed}$$

13.506 $$\frac{(3200 \text{ gpd})\ (8.34 \text{ lb/gal})\ (6.4/100)\ (68/100)\ (55/100)}{\frac{(0.785)\ (50 \text{ ft})\ (50 \text{ ft})\ (22 \text{ ft})}{1000}} = \frac{14.8 \text{ lb/d Destroyed}}{1000\text{ - cu ft}}$$

13.507 $$\frac{0.05 \text{ cfm}}{1 \text{ cu ft Digester Volume}} = \frac{\text{x cfm air required}}{(80 \text{ ft})\ (20 \text{ ft})\ (12 \text{ ft})}$$

(0.05) (80 ft) (20 ft) (12 ft) = x
x = 960 cfm Air

13.508 (22 mg/L) (0.12 MG) (8.34 lb/gal) = 22 lb Caustic

13.509 $$\frac{6.0 \text{ mg/L} - 3.8 \text{ mg/L}}{3 \text{ min}} \times 60 \text{ min/hr} = 44 \text{ mg/L/hr}$$

13.510 $$\frac{(119 \text{ mg/L})\ (2.2 \text{ MGD})\ (8.34 \text{ lb/gal})}{(3.0/100)} = (25 \text{ gpm})\ (8.34 \text{ lb/gal})\ (\text{x min/d})$$

$$\frac{(122)\ (2.4)\ (8.34)}{(0.030)\ (25)\ (8.34)} = x$$

390 min/d = x

$$\frac{390 \text{ min/d}}{24 \text{ hrs/d}} = 16.3 \text{ min/hr Pump Operating Time}$$

13.511 2.6% Solids: $\frac{(0.785)\ (32\text{ ft})\ (32\text{ ft})\ (24\text{ ft})\ (7.48\text{ gal/cu ft})}{12{,}000\text{ gpd}} = 12\text{ d}$

4.6% Solids: $\frac{(0.785)\ (32\text{ ft})\ (32\text{ ft})\ (24\text{ ft})\ (7.48\text{ gal/cu ft})}{5400\text{ gpd}} = 27\text{ d}$

13.512 $\frac{\frac{(1100\text{ gal})\ (8.34\text{ lb/gal})\ (3.8/100)}{3\text{ hr}}}{140\text{ sq ft}} = 0.83\text{ lb/hr/sq ft}$

13.513 $\frac{\frac{(820\text{ gal})\ (9.34\text{ lb/gal})\ (5/100)}{2\text{ hr}}}{160\text{ sq ft}} = 1.1\text{ lb/hr/sq ft}$

13.514 $\frac{(0.80\text{ lb/hr sq ft})\ (2\text{ hr})}{2\text{ hr} + 20\text{ min}}$

$= \frac{(0.80\text{ lb/hr sq ft})\ (2\text{ hr})}{2.33\text{ hr}}$

$= 0.7\text{ lb/hr/sq ft}$

13.515 $\frac{\frac{(680\text{ gal})\ (8.34\text{ lb/gal})\ (3.9/100)}{2\text{ hr}}}{130\text{ sq ft}} = 0.85\text{ lb/hr/sq ft}$

$\frac{(0.85\text{ lb/hr/sq ft})\ (2\text{ hrs})}{2.33\text{ hr}} = 0.7\text{ lb/hr/sq ft}$

13.516 $\frac{140\text{ gpm}}{6\text{ ft}} = 23\text{ gpm/ft}$

13.517 $\frac{21{,}300\text{ lb/d}}{12\text{ hr/d}} = 1775\text{ lb/hr}$

13.518 $1800\text{ lb/hr} = \frac{23{,}100\text{ lb/d}}{\text{x hr/d Operating Time}}$

$\text{x} = \frac{23{,}100\text{ lb/d}}{1800\text{ lb/hr}}$

$x = 13\text{ hr/d}$

13.519 (160 gpm) (60 min/hr) (8.34 lb/gal) (4.4/100) = 3,523 lb/hr

13.520 0.7% = 7,000 mg/L

$\frac{(4\text{ gpm})\ (1440\text{ min/day})}{1{,}000{,}000} = 0.00576\text{ MGD}$

$\frac{(7000\text{ mg/L})\ (0.00576\text{ MGD})\ (8.34\text{ lb/gal})}{24\text{ hr/d}} = 14\text{ lb/hr}$

13.521 $\frac{(80 \text{ gpm})(60 \text{ min/hr})(8.34 \text{ lb/gal})(5.1/100)}{320 \text{ sq ft}} = 6.4 \text{ lbs/hr/sq ft}$

13.522 $\frac{(6810 \text{ lb/hr})(31/100)}{320 \text{ sq ft}} = 6.6 \text{ lb/hr/sq ft}$

13.523 $3.3 \text{ lb/hr/sq ft} = \frac{\frac{5400 \text{ lb/d}}{\text{x hr/d operator}}}{230 \text{ sq ft}}(90/100)$

$3.3 \text{ lb/hr/sq ft} = \frac{5400 \text{ lb/d}}{\text{x hr/d}}(1/230 \text{ sq ft})(90/100)$

$\text{x} = \frac{(5400)(1)(90)}{(3.3)(230)(100)} = 6.4 \text{ hr/d operations}$

13.524 $\text{Filter Yield, lb/hr/sq ft} = \frac{\frac{18{,}310 \text{ lb/d}}{10 \text{ hr/d}}}{265 \text{ sq ft}} \times (91/100)$

$= \frac{1831 \text{ lb/hr}}{265 \text{ sq ft}}(91/100) = 6.3 \text{ lb/hr/sq ft}$

13.525 $\frac{(18{,}400 \text{ lb/hr})(20/100)}{(85{,}230 \text{ lb/hr})(5.9/100)} \times 100$

$\frac{3680 \text{ lb/hr}}{5029 \text{ lb/hr}} \times 100 = 73\% \text{ Solids Recovery}$

13.526 (210 ft) (22 ft) (0.67 ft) (7.48 gal/cu ft) = 23,154 gal

13.527 (240 ft) (26 ft) (0.67 ft) (7.48 gal/cu ft) = 31,272 gal

13.528 $\frac{\frac{(168{,}000 \text{ lb})(365 \text{ d/yr})(4.6/100)}{21 \text{ d}}}{(190 \text{ ft})(20\text{ft})} = 35.3 \text{ lb/yr/sq ft}$

13.529 (220 ft) (30 ft) (0.75) (7.48 gal/cu ft) (8.34 lb/gal) = 308,797 lb

$\frac{\frac{(308{,}797 \text{ lb})(365 \text{ d/yr})(39/100)}{25 \text{ d}}}{(220 \text{ ft})(30 \text{ ft})} = 26.6 \text{ lb/yr/sq ft}$

13.530 (0.785) (50 ft) (50 ft) (2.4 ft) = 4,710 cu ft

13.531 (0.785) (50 ft) (50 ft) (1.17 ft) = (70 ft) (40 ft) (x ft) = 0.82 ft

13.532 Sludge moisture is:

$\frac{(4700 \text{ lb/d})(79/100) + (3800 \text{ lb/d})(26/100)}{4700 \text{ lb/d} + 3800 \text{ lb/d}} \times 100$

$\frac{3713 \text{ lb/d} + 988 \text{ lb/d}}{8500 \text{ lb/d}} \times 100 = 55\%$

13.533 Sludge moisture = 83%.

$$42 = \frac{(4800\text{ lb/d})\ (83/100) + (\text{x lb/d})\ (27/100)}{(4800\text{ lb/d} + \text{x lb d})} \times 100$$

$$42/100 = \frac{3984\text{ lb/d} + (\text{x lb/d})\ (0.27)}{4800\text{ lb/d} + \text{x lb/d}}$$

$0.42\ (4{,}800 + x) = 3{,}984 + 0.27x$
$2{,}016 + 0.42x = 3{,}984 + 0.27x$
$0.42x - 0.27x = 3{,}984 - 2{,}016$
$0.15x = 1{,}968$
$x = 13{,}120$ lb/d

13.534

$$\frac{(7.4\text{ yds}^3\text{Sludge})\ (1710\text{ lb/yd}^3)\ (19/100) + (7.4\text{ yd}^3)\ (3)\ (760\text{ lb/yd}^3)\ (54/100)}{(7.4\text{ cu yds Sludge})\ (1710\text{ lb/cu yds}) + (7.4\text{ cu yds})\ (3)\ (760\text{ lb/cu yds})} \times 100$$

$$= \frac{2404\text{ lb Solids from Sludge} + 9111\text{ lb Solids from the wood chips}}{12{,}654\text{ lb Sludge} + 16{,}872\text{ lb wood chips}} \times 100$$

$$= \frac{11{,}515\text{ lb Solids}}{29{,}526\text{ lb Compost Blend}} \times 100$$

= 39% solids in compost blend

13.535

$$21\text{ days} = \frac{8200\text{ cu yds}}{\dfrac{\text{x lb/d}}{1000\text{ lb/cu yd}}}$$

$$21\text{ d} = \frac{(8200\text{ cu yds})\ (1000\text{ lb/cu yd})}{\text{x lb/day}}$$

$$\text{x lb/day} = \frac{(8200\text{ cu yds})\ (1000\text{ lb/cu yd}}{21\text{ d}}$$

$x = 390{,}476$ lb/d

13.536

$$\frac{(12\text{ yds}^3\text{ Sludge})\ (1720\text{ lb/yd}^3)\ (16/100) + (12\text{ yds}^3)\ (3)\ (820\text{ lb/yd}^3)\ (55/100)}{(12\text{ cu yds Sludge})\ (1720\text{ lb/cu yd}) + (12\text{ cu yds})\ (3)\ (820\text{ lb/cu yd})} \times 100$$

$$\frac{3302\text{ lb Solids from the Sludge} + 16{,}236\text{ lb Solids from the wood chips}}{20{,}640\text{ lb Sludge} + 29{,}520\text{ lb wood chips}} \times 100$$

$$\frac{19{,}538\text{ lb Solids}}{50{,}160\text{ lb Compost Blend}} \times 100 = 39\%\text{ Solids in Compost Blend}$$

13.537 $$21\text{ d} = \frac{(7810\text{ cu yds})\ (1100\text{ lb/cu yd})}{\frac{\text{x lb/d dry solids}}{0.19} + \frac{(\text{x lb/d dry solids})}{0.19}\ \frac{(3\text{ mix ratio})}{1}\ \frac{(780\text{ lb/cu yd})}{(1720\text{ lb/cu yd})}}$$

$$21 = \frac{7{,}810{,}000}{x/0.19 + 7.16x}$$

$$21 = \frac{7{,}810{,}000}{1/0.19x + 7.16x}$$

$$21 = \frac{7{,}810{,}000}{5.26x + 7.16x}$$

$$21 = \frac{7{,}810{,}000}{12.42x}$$

$$12.42x = \frac{7{,}810{,}000}{21}$$

$$x = \frac{7{,}810{,}000}{(21)\ (12.42)}$$

$x = 29{,}994$ lb/d dry solids

13.538 (150 gpm) (60 min/hr) (8.34 lb/gal) (4.8/100)= 3,603 lb/hr

13.539 (220 ft) (24 ft) (0.83) (7.48 gal/cu ft) = 32,780 gal

32,780 gal (3.3/100) (8.34 lb/gal) = 9,022 lb

$$\frac{9022\text{ lb}}{22\text{ d}} = 410.09\text{ lb/d}$$

(410.09 lb/d) (365 d/yr) = 149,683 lb/yr

(149,683 lb/yr) (1/(220 ft) (24 ft) = 28.3 lb/yr/sq ft

13.540 $$\frac{(0.20\text{ MG/day})\ (1{,}000{,}000\text{ gal})}{1440\text{ min/day}} = 139\text{ gpm}$$

13.541 (960 gal) (94.2/100) (8.34 lb/gal) = 336.27 lb

$$\frac{336.27\text{ lb}}{140\text{ min}}(60\text{ min/hr}) = 144\text{ lb/hr}$$

$$\frac{144\text{ lb/hr}}{150\text{ sq ft}} = 0.96\text{ lb/hr/sq ft}$$

13.542 Area = (3.14) (9.6 ft) (10 ft) = 301 sq ft

(36 gpm) (60 min/hr) (8.34 lb/gal) (12/100) = 2,162 lb/hr

$$\frac{2162\text{ lb/hr}}{301\text{ sq ft}} = 7.2\text{ lb/hr/sq ft}$$

13.543 (3,020 lb/hr) (40/100) = 1,208 lb/hr (24 gal/min) (60 min/hr) (8.50 lb/gal) (11/100) = 1,346 lb/hr

$$\%\text{ Recovery} = \frac{1208\text{ lb/hr}}{1346\text{ lb/hr}} \times 100 = 90\%$$

13.544 $$\frac{25{,}200\text{ lb/d}}{12\text{ hr/d}} = 2100\text{ lb/hr}$$

13.545 $\frac{(800 \text{ gal/2 hr})\ (8.34 \text{ lb/gal})\ (4.1/100)}{141 \text{ sq ft}} = 0.97 \text{ lb/hr/sq ft}$

13.546 (170 gpm) (60 min/hr) (8.34 lb/gal) (5/100) = 4,253 lb/hr

$$\frac{4253 \text{ lb/hr}}{2000 \text{ lb/ton}} = 2.1 \text{ tons/hr}$$

$$\frac{(2.8 \text{ gpm})\ (1440 \text{ min/day})}{1{,}000{,}000} 0.0040 \text{ MGD}$$

(9,000 mg/L) (0.0040 MGD) (8.34 lb/gal) = 300 lb/d

$$\frac{300 \text{ lb/d}}{24 \text{ hr/d}} = 12.5 \text{ lb/hr Flocculant}$$

$$\frac{12.5 \text{ lb/hr}}{2.1 \text{ tons/hr solids}} = 5.95 \text{ lb Flocculant/ton solids}$$

13.547 $(0.8 \text{ lb/hr/sq ft}) \frac{(2 \text{ hr})}{2 \text{ hr} + 20 \text{ minutes}}$

$$(0.8 \text{ lb/hr/sq ft}) \frac{(2 \text{ hr})}{2.33 \text{ hr}} = 0.69 \text{ lb/hr/sq ft}$$

13.548 24,300 mg/L – 740 mg/L = 23,560 mg/L

13.549 $\frac{(80 \text{ gpm})\ (60 \text{ min/hr})\ (8.34 \text{ lb/gal})\ (5.5/100)}{320 \text{ sq ft}} = 6.9 \text{ lb/hr/sq ft}$

13.550 $\frac{(7500 \text{ lb/hr } (26/100)}{320 \text{ sq ft}} = 6.1 \text{ lb/hr/sq ft}$

13.551 $1800 \text{ lb/hr} = \frac{28{,}300 \text{ lb/day}}{x \text{ hr/day}}$

$$x = \frac{28{,}300 \text{ lb/day}}{1800 \text{ lb/hr}} = 15.7 \text{ hr/day}$$

13.552 $3.1 = \frac{\frac{5700 \text{ lb/d}}{x \text{ hr/d}}}{280 \text{ sq ft}} \times (92/100)$

$$3.1 = \frac{5700 \text{ lb/d}}{x \text{ lb/d}} \times \frac{(1)}{280 \text{ sq ft}} \times \frac{(92)}{100}$$

$$x = \frac{(5700)\ (1)\ (2)}{(3.1)\ (280)\ (100)} = 6 \text{ hr/d operation}$$

13.553 (220 ft) (30 ft) (0.75 in) (7.48 gal/cu ft) = 37,026 gal

13.554 $\frac{(14{,}300 \text{ lb/hr})\ (28/100)}{91{,}000 \text{ lb/hr sludge } (5.3/100)} \times 100$

$$\frac{4004 \text{ lb/hr}}{4823 \text{ lb/hr}} \times 100 = 83\% \text{ solids recovery}$$

13.555 (8 in = 0.67 ft)(200 ft) (25 ft) (0.67 ft) (7.48 gal/cu ft) (8.34 lb/gal) = 208,984 lb sludge

$$\frac{\dfrac{(208{,}984\text{ lb})}{20\text{ d}}\dfrac{(365\text{ d})}{\text{yr}}(5.1/100)}{(200\text{ ft})\ (25\text{ ft})} = \frac{(183{,}070\text{ lb/yr})}{5000\text{ sq ft}} = 36.6\text{ lb/yr/sq ft}$$

13.556 $$\frac{(0.785)\ (40\text{ ft})\ (40\text{ ft})\ (1)}{(190\text{ ft})\ (30\text{ ft})} = \text{x}$$

$x = 0.22$ ft

13.557 Moisture content = 75%

$$55 = \frac{(6800\text{ lb/d})\ (75/100) + (\text{x lb/d})\ (36/100)}{6800\text{ lb/d} + \text{x lb/d}} \times 100$$

$$(55/100) = \frac{5100\text{ lb/d} + (\text{x lb/d})\ (0.36)}{6800\text{ lb/d} + \text{x lb/d}}$$

$0.55\ (6{,}800\text{ lb/d} + x) = 5{,}100 + 0.36x$

$3{,}740 + 0.55x = 5{,}100 + 0.36x$

$0.55x - 0.36x = 5{,}100 - 3{,}740$

$0.19x = 1{,}360$

$x = 7{,}158$ lb/d compost required

13.558 $$\frac{(7.0\text{ yd}^3\text{ Sludge})\ (1710\text{ lb/yd}^3)\ (16/100)}{(7.0\text{ cu yd Sludge})\ (1710\text{ lb/cu yd})} \quad \frac{+(7.0\text{ yd}^3)\ (3)\ (780\text{ lb/yd}^3)\ (51/100)}{+(7.0\text{ cu yds})\ (3)\ (780\text{ lb/cu yds})} \times 100$$

$$\frac{1915\text{ lb Solids from the Sludge} + 8354\text{ lb Solids from the wood chips}}{11{,}970\text{ lb Sludge} + 16{,}380\text{ lb wood chips}} \times 100$$

$$\frac{10{,}269\text{ lb Solids}}{28350\text{ lb Compost Blend}} \times 100 = 36.2\%\text{ Solids in Compost Blend}$$

13.559 $$26\text{ d} = \frac{6350\text{ cu yds}}{\dfrac{\text{x lb/d}}{980\text{ lb/cu ft}}}$$

$$26\text{ d} = \frac{6350\text{ cu yds} \times 980\text{ lb/cu yd}}{\text{x lb/day}}$$

$$\text{x} = \frac{6350\text{ cu yds} \times 980\text{ lb/cu yd}}{26\text{ d}}$$

$x = 239{,}346$ lb/d

$$\frac{239{,}346\text{ lb/d}}{2000\text{ lb/ton}} = 120\text{ tons/d}$$

13.560 $$24\text{ d} = \frac{9000\text{ cu yds} \times 1100\text{ lb/cu yd}}{\dfrac{\text{x lb/d}}{\text{Dry Solids}/0.18} + \dfrac{(\text{x lb/d})}{\text{Dry Solids}/0.18} \times \dfrac{3.3\text{ mix ratio}}{1} \times \dfrac{800\text{ lb/cu yd}}{1710\text{ lb/cu yd}}}$$

$$24 = \frac{9{,}900{,}000}{(1/0.18)x + (8.58x)}$$

$$24 = \frac{9{,}900{,}000}{5.56x + 8.58x}$$

$$24 = \frac{9{,}900{,}000}{14.14x}$$

$$14.14x = \frac{9{,}900{,}000}{24}$$

$$x = \frac{9{,}900{,}000}{24 \times 14.14}$$

x = 29,204 lb/day dry solids

Answers to Additional Practice Problems

1. (0.785) (70 ft) (70 ft) (25 ft) (7.48 gal/cu ft) = 719,295.5 gal
2. (60 ft) (20 ft) (10 ft) = 12,000 cu ft
3. (20 ft) (60 ft) (12 ft) (7.48 gal/cu ft) = 107,712 gal
4. (20 ft) (40 ft) (12 ft) (7.48 cu ft) = 71,808 gal
5. (0.785) (60 ft) (60 ft) (12 ft) (7.48 gal/cu ft) = 253,662 gal
6. (20 ft) (50 ft) (16 ft) (7.48 gal/cu ft) = 119,680 gal
7. (4 ft) (6 ft) (340 ft) = 8,160 cu ft
8. (0.785) (0.83 ft) (0.83 ft) (1,600 ft) (7.48 gal/cu ft) = 6,472 gal
9. 5 ft + 10 ft/2 (4 ft) (800 ft) (7.48 gal/cu ft)
   = (7.5 ft) (4 ft) (800 ft) (7.48 gal/cu ft)
   = 179,520 gal
10. (0.785) (.66) (.66) (2,250 ft) (7.48 gal/cu ft) = 5,755 gal
11. (5 ft) (4 ft) (1,200 ft) (7.48 gal/cu ft) = 179,520 gal
12. $\dfrac{(4\text{ ft})\ (4\text{ ft})\ (1200\text{ ft}}{27\text{ cu ft/cu yd}} = 711\text{ cu yds}$
13. (500 yd) (1 yd) (1.33 yd) = 665 cu yd
14. (900 ft) (3 ft) (3 ft) = 8,100 cu ft
15. (700 ft) (6.5 ft) (3.5 ft) = 15,925 cu ft
16. (0.785) (90 ft) (90 ft) (25 ft) (7.48 gal/cu ft) = 1,189,040 gal
17. (80 ft) (16 ft) (20 ft) = 25,600 cu ft
18. (0.785) (0.67 ft) (0.67 ft) (4,000 ft) (7.48 gal/cu ft) = 10,543 gal
19. (1,200 ft) (3 ft) (3 ft) = 10,800 cu ft

20. $\dfrac{(3\ ft)(4\ ft)(1200\ ft)}{27 \text{ cu ft/cu yd}} = 533 \text{ cu yds}$
21. (30 ft) (80 ft) (12 ft) (7.48 gal/cu ft) = 215,424 gal
22. (8 ft) (3.5 ft) (3,000 ft) (7.48 gal/cu ft) = 628,320 gal
23. (0.785) (70 ft) (70 ft) (19 ft) (7.48 gal/cu ft) = 546,665 gal
24. (0.785) (25 ft) (25 ft) (30 ft) (7.48 gal/cu ft) = 110,096 gal
25. (2.4 ft) (3.7 ft) (2.5 fps) (60 sec/min) = 1,332 cfm
26. (20 ft) (12 ft) (0.8 fpm) (7.48 gal/cu ft) = 1,436 gpm
27. $\dfrac{(4 \text{ ft} + 6 \text{ ft})}{2}(3.3 \text{ ft})\ (130 \text{ fpm})$
    = 5 (3.3 ft) (130 fpm)
    = 2,145 cfm
28. (0.785) (0.66) (0.66) (2.4 fps) (7.48 gal/cu ft) (60 sec/min) =368 gpm
29. (0.785) (3 ft) (3 ft) (4.7 fpm) (7.48 gal/cu ft) = 248 gpm
30. (0.785) (0.83 ft) (0.83 ft) (3.1 fps) (7.48 gal/cu ft) (60 sec/min) (0.5) = 376 gpm
31. (6 ft) (2.6 ft) (x fps) (60 sec/min) (7.48 gal/cu ft) = 14,200 gpm
    $x = 2.03$ ft
32. (0.785) (0.67) (0.67) (x fps) (7.48 gal/cu ft) (60 sec/min) = 584 gpm
    $x = 3.7$ fps
33. 550 ft/208 sec = 2.6 fps
34. (0.785) (0.83 ft) (0.83 ft) (2.4 fps) = (0.785) (0.67 ft) (0.67 ft) (x fps)
    $x = 3.7$ fps
35. 500 ft/92 sec = 5.4 fps
36. (0.785) (0.67) (0.67) (3.2 fps) = (0.785) (0.83 ft) (0.83 ft) (x fps)
    $x = 2.1$ fps
37. 35.3 MGD/7 = 5 MGD
38. 121.4 MG/30 days = 4.0 MGD
39. 1,000,000 × 0.165 = 165,000 gpd
40. 3,335,000 gal/1,440 min = 2,316 gpm
41. (8 cfs) (7.48 gal/cu ft) (60 sec/min) = 3,590 gpm
42. (35 gps) (60 sec/min) (1,440 min/day) = 3,024,000 gpd
43. $\dfrac{4{,}570{,}000 \text{ gpd}}{(1440 \text{ min/day})\ (7.48 \text{ gal/cu ft})} = 424 \text{ cfm}$
44. (6.6 MGD) (1.55 cfs/MGD) = 10.2 cfs
45. $\dfrac{(445{,}875 \text{ cfd})\ (7.48 \text{ gal/cu ft})}{1440 \text{ min/day}} = 2316 \text{ gpm}$
46. (2,450 gpm) (1,440 min/day) = 3,528,000 gpd
47. (6 ft) (2.5 ft) (x fps) (7.48 gal/cu ft) (60 sec/min) = 14,800 gpm
    $x = 2.2$ fps
48. (4.6 ft) (3.4 ft) (3.6 fps) (60 sec/min) = 3,378 cfm
49. 373.6/92 days = 4.1 MGD
50. (12 ft) (12 ft) (0.67 fpm) (7.48 gal/cu ft) = 722 gpm

51. (0.785) (0.67 ft) (0.67 ft) (x fps) (7.48 gal/cu ft) (60 sec/min) = 510 gpm
    x = 3.2 fps
52. (10 cfs) (7.48 gal/cu ft) (60 sec/min) = 4,488 gpm
53. 134.6/31 days = 4.3 MGD
54. (5.2 MGD) (1.55 cfs/MGD) = 8.1 cfs
55. (0.785) (2 ft) (2 ft) (3.3 fpm) (7.48 gal/cu ft) = 77.5 gpm
56. $\frac{(1,825,000 \text{ gpd})}{(1440 \text{ min/day})\ (7.48 \text{ gal/cu ft})} = 169 \text{ cfm}$
57. (0.785) (0.5 ft) (0.5 ft) (2.9 fps) (7.48 gal/cu ft) (60 sec/min) = 255 gpm
58. (0.785) (0.83 ft) (0.83 ft) (2.6 fps) = (0.785) (0.67 ft) (0.67 ft) (x fps)
    x = 4.0 fps
59. (2,225 gpm) (1,440 min/day) = 3,204,000 gpd
60. 5,350,000 gal/1,440 min/day = 3,715 gpm
61. (2.5 mg/L) (5.5 MGD) (8.34 lb/gal) = 115 lb/day
62. (7.1 mg/L) (4.2 MGD) (8.34 lb/gal) = 249 lb/day
63. (11.8 mg/L) (4.8 MGD) (8.34 lb/gal) = 472 lb/day
64. $\frac{(10 \text{ mg/L})\ (1.8 \text{ MGD})\ (8.34 \text{ lbs/gal})}{0.65} = 231 \text{ lbs/day}$
65. (41 mg/L) (6.25 MGD) (8.34 lb/gal) = 214 lb/day
66. (60 mg/L) (0.086 MGD) (8.34 lb/gal) = 43 lb
67. (2,220 mg/L) (0.225) (8.34 lb/gal) = 4,166 lb
68. $\frac{(8 \text{ mg/L})\ (0.83 \text{ MGD})\ (8.34 \text{ lbs/gal})}{0.65} = 85 \text{ lbs/day}$

Answers to Advanced Practice Problems

1. (0.785) (60 ft) (60 ft) (20 ft) (7.48 gal/cu ft) = 422,769.6 gal
2. (50 ft) (15 ft) (10 ft) = 7,500 cu ft
3. (10 ft) (50 ft) (10 ft) (7.48 gal/cu ft) = 37,400 gal
4. (10 ft) (40 ft) (10 ft) (7.48 cu ft) = 29,920 gal
5. (0.785) (50 ft) (50 ft) (12 ft) (7.48 gal/cu ft) = 176,154 gal
6. (20 ft) (40 ft) (16 ft) (7.48 gal/cu ft) = 95,744 gal
7. (4 ft) (6 ft) (300 ft) = 7,200 cu ft
8. (0.785) (0.83 ft) (0.83 ft) (1,500 ft) (7.48 gal/cu ft) = 6,068 gal
9. 5 ft + 10 ft/2 (4 ft) (700 ft) (7.48 gal/cu ft)
   = (7.5 ft) (4 ft) (700 ft) (7.48 gal/cu ft)
   = 157,080 gal
10. (0.785) (0.66) (0.66) (2,100 ft) (7.48 gal/cu ft) = 5,371 gal
11. (5 ft) (4 ft) (1,100 ft) (7.48 gal/cu ft) = 164,560 gal
12. $\frac{(4\ ft)(4\ ft)(1100\ ft)}{27 \text{ cu ft/cu yd}} = 652 \text{ cu yds}$
13. (400 yd) (1 yd) (1.33 yd) = 532 cu yd
14. (810 ft) (3 ft) (3 ft) = 7,290 cu ft

15. (600 ft) (6.5 ft) (3.5 ft) = 13,650 cu ft
16. (0.785) (90 ft) (90 ft) (20 ft) (7.48 gal/cu ft) = 951,232 gal
17. (80 ft) (12 ft) (20 ft) = 19,200 cu ft
18. (0.785) (0.67 ft) (0.67 ft) (3,000 ft) (7.48 gal/cu ft) = 7,908 gal
19. (1,500 ft) (3 ft) (3 ft) = 13,500 cu ft
20. $\frac{(3\ ft)(4\ ft)(1100\ ft)}{27\ \text{cu ft/cu yd}} = 489\ \text{cu yds}$
21. (30 ft) (60 ft) (12 ft) (7.48 gal/cu ft) = 161,568 gal
22. (8 ft) (3.5 ft) (2,000 ft) (7.48 gal/cu ft) = 418,880 gal
23. (0.785) (60 ft) (60 ft) (19 ft) (7.48 gal/cu ft) = 401,631 gal
24. (0.785) (20 ft) (20 ft) (30 ft) (7.48 gal/cu ft) = 70,462 gal
25. (2.4 ft) (3.7 ft) (2.0 fps) (60 sec/min) = 1,066 cfm
26. (20 ft) (12 ft) (0.7 fpm) (7.48 gal/cu ft) = 1,257 gpm
27. $\frac{(4\ \text{ft} + 6\ \text{ft})}{2}(3.3\ \text{ft})\ (120\ \text{fpm})$
    = 5 (3.3 ft) (120 fpm)
    = 1,980 cfm
28. (0.785) (0.66) (0.66) (2.2 fps) (7.48 gal/cu ft) (60 sec/min) =338 gpm
29. (0.785) (2 ft) (2 ft) (4.7 fpm) (7.48 gal/cu ft) = 110 gpm
30. (0.785) (0.83 ft) (0.83 ft) (3.0 fps) (7.48 gal/cu ft) (60 sec/min) (0.5) = 364 gpm
31. (6 ft) (2.5 ft) (x fps) (60 sec/min) (7.48 gal/cu ft) = 14,200 gpm
    x = 2.11 ft
32. (0.785) (0.67) (0.67) (x fps) (7.48 gal/cu ft) (60 sec/min) = 590 gpm
    x = 3.7 fps
33. (0.785) (0.83 ft) (0.83 ft) (2.4 fps) = (0.785) (0.67 ft) (0.67 ft) (x fps)
    x = 3.7 fps
34. 400 ft/92 sec = 4.3 fps
35. (0.785) (0.67) (0.67) (3.2 fps) = (0.785) (0.83 ft) (0.83 ft) (x fps)
    x = 2.1 fps
36. 35.3 MGD/7 = 5 MGD
37. 124.4 MG/30 days = 4.1 MGD
38. 1,000,000 × 0.175 = 175,000 gpd
39. 3,330,000 gal/1,440 min = 2,313 gpm
40. (7 cfs) (7.48 gal/cu ft) (60 sec/min) = 3,142 gpm
41. (30 gps) (60 sec/min) (1,440 min/day) = 2,592,000 gpd
42. $\frac{4{,}500{,}000\ \text{gpd}}{(1440\ \text{min/day})\ (7.48\ \text{gal/cu ft})} = 418\ \text{cfm}$
43. (6.5 MGD) (1.55 cfs/MGD) = 10.1 cfs
44. $\frac{(445{,}870\ \text{cfd})\ (7.48\ \text{gal/cu ft})}{1440\ \text{min/day}} = 2316\ \text{gpm}$
45. (2,400 gpm) (1,440 min/day) = 3,456,000 gpd
46. (6 ft) (2.0 ft) (x fps) (7.48 gal/cu ft) (60 sec/min) = 14,800 gpm
    x = 2.7 fps

47. (4.6 ft) (3.4 ft) (3.5 fps) (60 sec/min) = 3,284 cfm
48. 378.6/92 days = 4.1 MGD
49. (10 ft) (10 ft) (0.67 fpm) (7.48 gal/cu ft) = 501 gpm
50. (0.785) (0.67 ft) (0.67 ft) (x fps) (7.48 gal/cu ft) (60 sec/min) = 510 gpm
    $x = 3.2$ fps
51. (11 cfs) (7.48 gal/cu ft) (60 sec/min) = 4,937 gpm
52. 134.6/30 days = 4.5 MGD
53. (5 MGD) (1.55 cfs/MGD) = 7.8 cfs
54. (0.785) (2 ft) (2 ft) (3.3 fpm) (7.48 gal/cu ft) = 77.5 gpm
55. $\dfrac{(1{,}820{,}000\ gpd)}{(1440\ \text{min/day})\ (7.48\ \text{gal/cu ft})} = 169\ \text{cfm}$
56. (0.785) (0.5 ft) (0.5 ft) (2.7 fps) (7.48 gal/cu ft) (60 sec/min) = 238 gpm
57. (0.785) (0.83 ft) (0.83 ft) (2.6 fps) = (0.785) (0.67 ft) (0.67 ft) (x fps)
    $x = 4.0$ fps
58. (2,220 gpm) (1,440 min/day) = 3,196,800 gpd
59. 5,300,000 gal/1,440 min/day = 3,681 gpm

Appendix B: Formulae

1. Area
   a. Rectangular Tank
      $A = L \times W$
   b. Circular Tank
      $A = \pi r^2$ or $A = 0.785\ d^2$
2. Volume
   a. Rectangular Tank
      $V = L \times W \times H$
   b. Circular Tank
      $V = \pi r^2 \times H$ or $0.785\ d^2 \times H$
3. Flow
   gal/day (gpd) = gal/min (gpm) × 1,440 min/day
   gal/day (gpd) = gal/hr (gph) × 24 hr/day
   million gallons/day (MGD) = (gal/day)/1,000,000
4. Dose
   lb = ppm × MG × 8.34 lb/gal
   ppm = lb/(MG × 8.34 lb/gal)
5. Efficiency (% removal = $\dfrac{\textbf{(Influent - effluent)}}{\text{Influent}} \times 100$
6. Weir loading (overflow rate) = $\dfrac{\text{total gallons/d}}{\text{length of weir}}$
7. Surface settling rate = $\dfrac{\text{total gallons/d}}{\text{surface area of tank}}$
8. Detention time (hours) = $\dfrac{\text{capacity of tank (gal)} \times 24\ \text{hr/d}}{\text{flow rate (gal/d)}}$
9. Horsepower (Hp) = $\dfrac{\text{gpm} \times \text{head (ft)}}{3960 \times \text{total efficiency}}$

Index

F

G

H

I

J

K

L

M

Printed in the USA
CPSIA information can be obtained
at www.ICGtesting.com
LVHW010859211123
764459LV00003B/11